G000118178

Demographic Change and Housing Wealth

John Doling • Marja Elsinga

Demographic Change and Housing Wealth

Homeowners, Pensions and Asset-based Welfare in Europe

With Contribution by

Kees Dol
Delft University of Technology, The Netherlands

József Hegedüs
Metropolitan Research Institute, Budapest, Hungary

Nick Horsewood
University of Birmingham, UK

Deborah Quilgars
University of York, York, UK

Richard Ronald
University of Amsterdam, The Netherlands

Hanna Szemzo
Metropolitan Research Institute, Budapest, Hungary

Nóra Teller
Metropolitan Research Institute, Budapest, Hungary

Janneke Toussaint
Technische Universiteit Delft, The Netherlands

 Springer

John Doling
School of Social Policy
University of Birmingham
B15 2TT Birmingham
United Kingdom

Marja Elsinga
Research Institute OTB
Delft University of Technology
2628 BX Delft
The Netherlands

ISBN 978-94-007-4383-0   ISBN 978-94-007-4384-7 (eBook)
DOI 10.1007/978-94-007-4384-7
Springer Dordrecht Heidelberg New York London

Library of Congress Control Number: 2012944993

© Springer Science+Business Media Dordrecht 2013
This work is subject to copyright. All rights are reserved by the Publisher, whether the whole or part of the material is concerned, specifically the rights of translation, reprinting, reuse of illustrations, recitation, broadcasting, reproduction on microfilms or in any other physical way, and transmission or information storage and retrieval, electronic adaptation, computer software, or by similar or dissimilar methodology now known or hereafter developed. Exempted from this legal reservation are brief excerpts in connection with reviews or scholarly analysis or material supplied specifically for the purpose of being entered and executed on a computer system, for exclusive use by the purchaser of the work. Duplication of this publication or parts thereof is permitted only under the provisions of the Copyright Law of the Publisher's location, in its current version, and permission for use must always be obtained from Springer. Permissions for use may be obtained through RightsLink at the Copyright Clearance Center. Violations are liable to prosecution under the respective Copyright Law.
The use of general descriptive names, registered names, trademarks, service marks, etc. in this publication does not imply, even in the absence of a specific statement, that such names are exempt from the relevant protective laws and regulations and therefore free for general use.
While the advice and information in this book are believed to be true and accurate at the date of publication, neither the authors nor the editors nor the publisher can accept any legal responsibility for any errors or omissions that may be made. The publisher makes no warranty, express or implied, with respect to the material contained herein.

Printed on acid-free paper

Springer is part of Springer Science+Business Media (www.springer.com)

Foreword

The ageing of populations and the growth of individually owned housing assets are macro-processes of change that, during recent decades, have been common across what are now the member states of the European Union. Of the two processes, the former has probably attracted far more policy-related interest, being both a cause for celebration of longer life expectancies as well as a cause for concern in the fiscal challenges of meeting the pension and social care needs of an increasingly large retirement population. Certainly, what has come to be known as the 'pension crisis' has generated many hours of debate and many pages of report; currently, there is at least some agreement across member states about policy objectives in which the elongation of working lives has been central.

An almost entirely separate set of policy issues has related to the second process of change, the growth in the size of homeownership sectors in most of the member states, to the point where some two-thirds of European households now own their homes. Partly because of the tendency for house prices to increase in the long run, housing has become the single largest item in the aggregate wealth holdings of European households. The aspiration of people to share in the wealth potential of homeownership and the possibilities that the wealth provides have a significance in the social and economic lives of European citizens.

The coincidence of the two processes suggests the intriguing question of the extent to which homeownership provides a potential cure for some of the consequences of ageing populations, as well as contributes to the causes. More specifically, housing had become an increasingly important element in the composition of household wealth across EU member states. Currently, housing equity considerably exceeds total GDP, offering the opportunity for housing to be viewed – by households, by governments and by financial institutions – as an asset that might be realised in order to meet consumption needs of older people, needs that have gained added poignancy as the nature and extent of the 'pension crisis' and the public expenditure implications of paying for health and social care become clearer. In that way, housing wealth may be seen as a potential solution for challenges later on in the life cycle.

Any such solution is influenced by challenges and decision earlier in the life cycle. The high costs of entry into homeownership, particularly where the wealth potential of doing so is great and where there are few alternative options facing young people, appear to have placed added pressure on the need for separate households to have two, full-time incomes. One possibility is that young adults are being *forced*, by the high price of homeownership and limited alternative housing possibilities, to invest more, relative to consumption, than they would otherwise choose to do. This front-loading of their investments may lead to a number of compensatory strategies which include reducing investment in other forms of pension provision and reducing the number of children they have. Reaching old age, however, the majority of Europeans, by virtue of investment in homeownership earlier in their lives, have a financial asset that can in principle contribute to their consumption needs.

While in the past, there has not been widespread realisation of housing wealth for this purpose, a number of current developments are leading to changes. There are also many important scientific and policy questions. For households, these concern the way they view housing in the context of other forms of saving; their willingness to use housing assets as a pension, rather than say to leave them as a bequest for children; and how attitudes and behaviour might be changing in line with wider demographic changes that are resulting in people living for more years post retirement and having fewer children. From the perspective of governments, the issues concern whether, in the context of the challenges posed by demographic change, they see housing wealth as offering an alternative (and whether as a substitute or a complement) to social provision, met through taxation. If they do, what are the consequences of such an approach to welfare, for example, for younger people trying to enter homeownership and for those who will never be able to enter: in short, how does housing equity measure up against the usual expectations about the performance of pension systems? For their part, the questions related to financial institutions include how they have, and will, respond in the form of providing and marketing products that will facilitate accessing housing wealth.

The present book broadly responds to these questions, doing so by drawing on research undertaken as part of the DEMHOW (Demographic Change and Housing Wealth) project. Originally submitted as a proposal to the EU under its 7th Framework Programme early in 2006, the decision to provide funding for a 33-month project allowed a start date of 1 March 2007. The overall aim of DEMHOW was to investigate the ways in which, across member states, demographic change and housing wealth are linked and to use those investigations in order to contribute to policy making. The specific focus of the present book is on a large but nevertheless specific aspect, namely, the way in which homeownership has in the past and could in the future contribute to the income needs of older people.

Undertaken by a consortium of 12 partner institutions located in 10 different member states (see Box I for a full list of participants), DEMHOW has involved a range of research objectives and methodologies including both quantitative and qualitative approaches to understanding past behaviour and present attitudes, and the establishing of socio-economic relationships and policy analysis. These studies have been selectively brought together in the present book in a number of stages. The reports and the publications produced by the members of the full DEMHOW team were reviewed by members of four of the partner institutions: the Universities of Birmingham and York, the Delft Technical University and the Metropolitan Research Institute, Budapest, and specifically the individuals listed on the front cover of the present book, to produce a draft manuscript that was delivered to the Commission as one of the products of the project. Following discussion among the authors, it was decided to proceed to a final manuscript version by two of that list working further with the draft, omitting some, adding new parts and re-arranging most of the existing contents. Almost all chapters in the present version, therefore, are complex amalgamations of the chapters in the earlier draft, so that in most cases the contribution of individuals, beyond the two principal authors, cannot be simply identified. The book thus draws together a diverse body of researcher contributions and methodologies. We hope that we have done justice to this combined effort.

Birmingham   John Doling
Delft   Marja Elsinga

Box I DEMHOW partners and participants

Institution	Researchers
University of Birmingham, UK	John Doling
	Nick Horsewood
University of Ghent, Belgium	Pascal de Decker
	Vicky Palmans
University of Southern Denmark, Denmark	Jorgen Lauridsen
	Morten Skak
University of Turku, Finland	Paivi Naumanen
	Hannu Rounavaara
L'Agence Nationale pour l'Information sur le Logement, France	Bernard Vorms
University of Bremen, then Humboldt University, Berlin, Germany	Tim Geilenkeuser
	Ilse Helbrecht
Metropolitan Research Institute, Hungary	József Hegedüs
	Hanna Szemzo
	Nóra Teller
Delft University of Technology, the Netherlands	Kees Dol
	Marja Elsinga
	Peter Neuteboom
	Richard Ronald
	Janneke Toussaint
Centro de Estudas para a Intervencaon Social, Portugal	Pedro Perista
University of Ljubljana, Slovenia	Srna Mandic
University of York, UK	Mark Bevan
	Anwen Jones
	Deborah Quilgars
	Mark Stephens
AGE Platform Europe	Anne-Sophie Parent

Acknowledgements

We are grateful to the European Commission which together with our individual institutions provided funds that have enabled the *Demographic Change and Housing Wealth (DEMHOW: FP7-SSH-2007-216865)* research project to be undertaken.

Dominik Sobzcak, DG Research, has been our project officer throughout the project both providing a link with the Commission and helpfully supporting our research and dissemination activities.

The project also benefited from the input of an advisory group, chaired by Anne-Sophie Parent of the AGE Platform Europe, which was one of the project partners, with a membership consisting of Sander Scheurwater (RICS Europe), Alessandro Sciamarelli (European Mortgage Federation) and David Taylor and Christophe André (OECD).

Project funded under the Socio-economic Sciences and Humanities

Contents

Chapter 1
Issues and Approaches

1.1 Introduction

The development of an ageing Europe is associated with a number of policy issues. Central, given relatively shrinking working populations, is the fiscal challenge of meeting the pension entitlements accrued under existing commitments in different member states. Quite simply, how will these entitlements be met? This challenge is set alongside another development: the long-term growth of the value of assets in the form of owner-occupied housing built up by European households and leading to older Europeans having a large proportion of their total personal wealth held in the form of housing. The co-relating of these two developments is not gratuitous, particularly since in some member states there has been explicit recognition that housing wealth could, sometimes should, be considered as part of the answer to meeting the income needs of older people.

Whereas such recognition is undoubtedly contentious, it is at least consistent with what might be viewed as a general trend over recent decades in many parts of Europe and beyond. Everywhere, welfare states appear to be under threat (Pierson 2002; Frericks 2010). Not only are demographic trends – ageing as well as migration – seen to be shifting the demands placed on welfare services, but globalisation, a neoliberal hegemony and the current economic and financial crisis all appear to be restructuring the role of the state. Frequently portrayed as part of this is a shift from collective responsibility for meeting individual welfare needs, often through social protection measure in the form of transfer payments, towards a greater emphasis on personal responsibility and solutions based on the prior accumulation of personal financial assets: sometimes called asset-based welfare, which, in the present context focusing on homeownership, might be rephrased as property asset-based welfare.

Housing has, in any case, a very special place in welfare debates, with social housing often being described as the wobbly pillar of the welfare state (Torgerson 1987; Malpass 2008). Further, housing, and in particular homeownership, has

J. Doling and M. Elsinga, *Demographic Change and Housing Wealth: Homeowners, Pensions and Asset-based Welfare in Europe*, DOI 10.1007/978-94-007-4384-7_1,
© Springer Science+Business Media Dordrecht 2013

played a role in the family's financial strategy, particularly in countries where a collective welfare state has not been well developed (Poggio 2008). Arguably, then, both social housing and homeownership have played a significant role in security for old age.

Nevertheless, given the recognition in policy debates that housing wealth might be systematically utilised with a view to meeting income needs in older age, a raft of questions are begged. Do European households share the views expressed in some policy debates; what do Europeans think about using their housing assets in this way; why have they accumulated so much housing wealth in the first place and to what extent is it intended for pension and non-pension purposes; how could housing assets be realised; and if they were to be realised what could be their role in pension systems? These can be reduced to three general research questions, which together form the focus of the present book:

1. *The past behaviour question*: how have Europeans used housing assets in the past? The objective is to establish the extent to which households in different member states have actually accumulated and used housing assets, and especially how that use has extended to contributing to income needs in old age.
2. *The present attitudes question*: what are their current views and future expectations about how they should be used? Against the background of revealed behaviour, how do European households view housing equity in their life strategies and what are the similarities and differences between countries and generations?
3. *The policy outcome question*: how would housing assets perform as pensions? In other words, if housing assets were systematically used in order to meet income needs in old age, how would they measure up to standard criteria against which pension systems are generally assessed?

Insofar as these are questions about housing assets or equity and the use to which they are or might be put, they can be investigated empirically, for which our starting point is the prior question: why do households accumulate assets, that is save, at all? Our perspective on this is informed by Modigliani's life cycle model (LCM), which, in reduced form, posits that people save during their working years in order to enable them to consume during their retirement years; the LCM, thus, constitutes a process of consumption smoothing across the life cycle. In practice, this is frequently achieved by the accumulation, while working, of wealth in different forms – from saving in private pensions to collecting works of art, buying shares and acquiring housing. All of these, then, are investment vehicles that facilitate horizontal distribution across the life cycle.

Important to these possibilities, both to the means of accumulating assets in the first place and being able subsequently to realise them, is the development of financial markets. These enable households to consume and invest, in advance of the prior saving from income, while also allowing them later to realise accumulated investments in order to consume. Financial products have thus become integral to decisions about saving, dissaving, and consuming.

Studies testing the LCM have concluded that in addition to life cycle motives, those pertaining to precautionary motives – saving against accident or emergency – are also important, as too are bequest motives – the desire to pass on wealth to subsequent generations. Both suggest that examination of the scale of saving and dissaving behaviour, and cross country variations, needs to be extended. The first extension is to welfare systems. In all European member states, governments provide forms of social insurance or protection. These include pensions but also other transfer payments as well as direct provisions to cover eventualities such as incapacity through ill-health or unemployment. Insofar as government welfare provisions are financed by taxation, paying taxes during working years acts in a similar way to saving of private assets: both private, life cycle saving and tax-based welfare systems are forms of horizontal distribution over the life course. The propensity to accumulate individual assets might thus be expected to be lower in countries in which individuals are eligible to receive generous and comprehensive protection from welfare arrangements.

The second extension is to the family. In those countries in which the extended family forms a robust system for the support of its members, with transfers taking place across them, and across generations, it too acts as a form of horizontal distribution across the life course. In effect, family members may contribute cash to the family economy during periods when they are in paid employment, and also contribute other activities such as care when they are able. The transfer of the family home from one generation to another can be seen, then, as a form of inheritance that helps to secure the family project. This appears very important in southern and eastern Europe (Poggio 2012; Mandic 2012), but Mulder (2007) suggests the importance is also increasing in other parts of Europe for three reasons: there are fewer children and only one or two siblings who have to share the inheritance; there are more immigrants for whom it is normal to rely more on family networks; and there are more divorces so that wider family networks may become more important in the broad sense.

In the following sections of this chapter, a number of the building blocks developing these initial points are presented in rather more detail. The first of these is the policy agenda which, shaped by demographic and housing sector developments, has led to some exploration – European-wide as well as member state specific – of the possibilities of using housing equity to meet income needs. The second presents a context for our three main questions – how have housing assets been used in the past; how are they viewed at the present time; and how might they perform as a pension in future. This starts from consideration of how housing assets fits into a wider framework of motives for which, and of means by which, households save and spend throughout the life course. This leads into presentation of the methodologies adopted in order to address our three main questions. The final section of the chapter presents the books content and structure through a chapter by chapter development of the findings and arguments.

1.2 Demographic and Housing Developments: Policy Challenges

1.2.1 Demographic Change

The ageing of Europe can be seen as part of a general, world-wide trend, taking place over the course of more than a century and through which there has been a shift from high to low rates of fertility and mortality (Kinsella and Phillips 2005). This demographic transition is bringing to an end a historically short period – much of the middle and latter part of the twentieth century – when many European, as well as other western economies, had, in economic development terms, an unusually favourable age structure. From the 1930s, birth rates declined, thus moving away from the youth dependency of the past. For their part, mortality rates had not yet reduced significantly so that retired people were relatively few in number, and age dependency was yet to emerge. In contrast, the present period is one in which 'that benevolent phase of population structure, a transitional phase between the youth dependency of the past and the aged dependency of the future, is now going' (Coleman 2001: 2).

There are detailed differences between member states in the extent of these trends. The total fertility rate, or average number of expected births in a woman's lifetime, has dropped everywhere, but to particularly low levels in the southern and some eastern member states (Castles 2004; European Commission 2005a). Yet, overall the picture is uniform: throughout Europe the total fertility rate is below replacement levels. Likewise, life expectancies have increased. Existing forecasts of both trends are that they will continue. The consequences for the statistical relationship between age groups are dramatic. In 1950, the countries that were to become the EU25, had, on average, only 9.1% of their populations aged 65 and over, with 24.9% under 15, whereas the forecast for a century later shows a reversal to 30.4% and 13.3%, respectively (Table 1.1). Correspondingly, while for every one person of retirement age in 1950 there were 5.52% younger adults aged 25–64 years, the expectation is that by 2050 there will be only 1.52%.

Table 1.1 Distribution of the population (EU25) by age group

	1950	1975	2000	2025	2050
80+	1.2	2.0	3.4	6.5	11.9
65–79	7.9	10.7	12.3	16.2	18.5
50–64	15.2	15.4	17.2	21.3	18.5
25–49	35.0	32.7	36.9	31.1	28.2
15–24	15.8	15.5	13.0	10.5	9.7
0–14	24.9	23.7	17.1	14.4	13.3
Ratio of older (>65):younger (24–64) adults	1:5.52	1:3.79	1:2.85	1:2.31	1:1.52
Median age	31	33	39	45	48

Source: European Commission (2005a)

This changing balance between age groups is unevenly affected by another demographic trend: migration both between the member states and between the EU as a whole and the rest of the world. The precise numbers involved are difficult to estimate, especially with illegal migration playing an important part. According to the Demography Report (2008), however, the annual number of international migrants in the last two decades of the twentieth century was around half a million, while since 2002, net migration into the EU has more than tripled to between 1.6 and 2 million people per year. One of the characteristics of migrants across international borders is that they tend to be relatively young, often being young adults, frequently with children, and in that way they may provide receiving countries with a counterbalance to ageing trends.

Detailed comparison within the European Union shows that the demographic changes have a very different effect on the constituent countries, creating an asymmetrical demographic pressure. Taking both the differences in fertility rate and the migratory patterns together, it is possible to differentiate three major areas of demographic change within the confines of the European Union (Sobotka 2008):

- *Southern Europe and the German speaking countries.* In these areas, low fertility is combined with replacement migration that can help to compensate the population reduction. Here, Germany is something of an exception as a consequence of East German population trends.
- *Central and Eastern Europe.* Low levels of fertility combined with very little or no migration at all, or in a few cases extensive emigration. These areas are already experiencing population loss, particularly of younger, working age people.
- *Western and northern countries.* This is the relatively high fertility belt of Europe, where there is also a high level of migration. Here, the predictions show a fairly stable or even growing population well into the middle of the twenty-first century.

This asymmetrical demographic change will mean an asymmetrical fiscal pressure in the countries of the EU, especially since it will have associated impacts on economic prosperity. These are also considerable. The changes that have occurred already, as well as those yet to come, present significant challenges to member states in terms both of reducing economic growth potential below what it might otherwise be and of meeting the funding of health care and pension needs (European Communities 2004; Malmberg 2007). In the mid 2000s, the Kok report concluded that

> these developments will have profound implications for the European economy and its capability to finance European welfare systems…the pure impact of ageing populations will be to reduce the potential growth rate…. a GDP per head some 20% lower than could otherwise be expected…. [and] an increase in pension and healthcare spending by 2050, varying between 4% and 8% of GDP (European Communities 2004:13).

Many current forecasts suggested that the consequences, if anything, have worsened since the publication of the Kok report, not least because of the impact of the economic and financial crisis of the last years of the decade. Subsequent EU Green Papers (European Commission 2005a, 2010a, b) have repeated the concern about the demographic trend and the scale of the challenge, as has the recent strategy paper, mapping out Europe's strategic objectives over the next decade (European

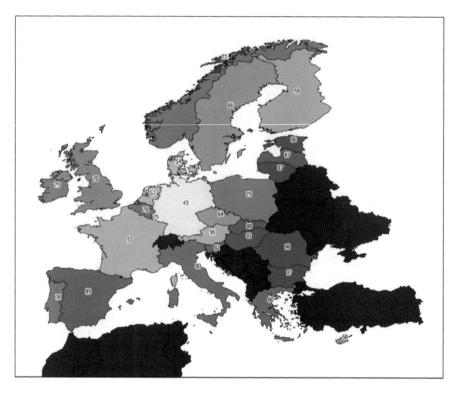

Fig. 1.1 Homeownership rate late 2000s (Source: European Mortgage Federation, National Statistical Institutes)

Commission 2010a). In setting out the main structural weaknesses of the EU, this summarised the mainstream view:

> Demographic ageing is accelerating. As the baby-boom generation retires, the EU's active population will start to shrink as from 2013 to 2014. The number of people aged over 60 is now increasing twice as fast as it did before 2007 – by about two million every year compared to one million previously. The combination of a smaller working population and a higher share of retired people will place additional strains on our welfare systems (European Commission 2010a: 5).

1.2.2 An Increasing Number of Homeowners

In 1945, homeownership was a minority tenure in each of what became the EU 25. By 2003, homeownership was the majority tenure in every country except Germany (Fig. 1.1).

However, the exact enumeration of homeowners in each country, and hence comparison between countries is sometimes difficult largely due to the absence of

Table 1.2 Homeownership in new and old member states (c 2003)

	Total housing stock	Homeownership rate	Total homeownership stock
	(1,000s)	(%)	(1,000s)
EU15	178,420	63.5	113,351.0
EU10	26,243	66.7	17,494.9
EU25	204,663	63.9	130,845.9

Source: Doling and Ford (2007)

systematically collected and fully harmonised data. Surveys of the housing stock by tenure have not been carried out at regular periods in each country while definitions of what constitutes homeownership differ from country to country. For example, Stephens points out that Swedish cooperative housing often does not appear in the official statistics as part of the homeownership sector; however, because it is a tradable asset, it could be thought of as such, and were it to be re-allocated, it would significantly raise the official homeownership level in that country (Stephens 2003). Nevertheless, taking national definitions at face value, the homeownership rate across the 15 pre-2004 member states had by 2003 reached 63.5%, and in the 10 newer member states, 66.7% (Table 1.2). In the light of this, it may be appropriate to refer to Europe as a *Union of Homeowners* (Doling and Ford 2007).

Explanations of the growth in homeownership across Europe have largely focused on particular economic and political developments at the macro level and have included the impact of rising incomes and affluence following post-war re-construction in a number of countries; favourable fiscal treatment (including tax benefits); rising demand and the hedge against inflation provided by property ownership; and direct policy intervention such as the legislation in the UK that gave tenants the right to buy at a reduced price and the various forms of transfer of rental property to owned property seen in some of the transition countries. At a general level, emphasis has also been placed on the growing impact of globalisation and the influence of resurgent neo-liberal ideologies (Doling and Ford 2003).

Notwithstanding the apparent universality of these trends, and probably therefore also the apparent cross country similarities, the meaning of homeownership and indeed other tenures varies across countries (Mandic and Clapham 1996; Ruonavaara 1993; Elsinga and Hoekstra 2005). In some northwestern European countries, for example, the Netherlands and Denmark, renting is generally an acceptable alternative to homeownership, providing good quality housing, security of tenure and attracting a positive image. The welfare state in these countries is also generous. This may contribute to understanding why the homeownership rate is relatively low in these countries.

Whatever the explanations for the growing dominance of homeownership, this is a dominance of homeownership as not only providing somewhere to live but also as a financial asset. The growth in the numbers of homeowners has in recent decades been accompanied by a general growth in house prices. In addition, while more Europeans are borrowing more money from financial institutions in order to enter homeownership and later to move to more expensive housing, in general they also pay off their housing loans. One outcome is an ever-growing amount of housing equity owned outright by individual households. Actually, the statistical data on this are also poor but rough estimates suggest that across the EU, total net housing equity may equate to about 140% of GDP (Doling and Ford 2007). In southern and eastern Europe, it may amount to 200–300% of GDP.

1.2.3   Housing Asset-Based Welfare

The recognition of a similarity in ageing and homeownership trends does not of itself establish anything other than coincidence. The basis of our research has been that there is active policy interest in the perceived potential of the use of housing assets to counter some of the adverse consequences of ageing and this can be recognised in the substance of academic and policy debates. Central to these have been analyses that see western welfare states confronted by permanent austerity: globalisation, economic slowdown and crisis, and ageing, all of which seem unlikely to diminish in intensity and all of which create fiscal stress to which welfare states are forced to react (Pierson 2001). While some versions of this argument stress the uniformity of national responses, others argue that different welfare regimes generate quite different political dynamics of welfare re-structuring (Pierson 2002; Castles 2004). As Starke et al. (2008) conclude, welfare states in the OECD countries may be under reconstruction but they are not necessarily converging.

1.2.3.1   Asset-Based Welfare

Converging or not, one dimension of the restructuring, which has figured prominently in some countries though hardly at all in others, is the notion of asset-based welfare. The principle underlying this is that, rather than relying on state-managed social transfers to counter the risks of poverty, individuals should be encouraged, and enabled, to accept greater responsibility for their own welfare needs by investing in financial products and property assets which augment in value over time. These can, at least in theory, later be tapped to supplement consumption and welfare needs when income is reduced, for example, in retirement, or used to acquire other forms of investment such as educational qualifications. Sherraden argues that whereas income transfers enable consumption over the next period, assets, in contrast, free up people to pursue long-term goals to enable individuals to make substantive and life-changing decisions, for example, to set up a small business or to undertake training. In such ways the individual gains by becoming self-reliant, the tax payer gains through reductions in the need for continued state benefit payments and the economy gains through additional participation in the labour force (Sherraden 1991, 2003).

Some countries have developed initiatives consistent with this approach. In the USA, Individual Development Accounts, introduced in 1997, encouraged lower income groups to save by matching their own contributions with savings provided from public funds (McKay 2002). In Sweden, some firms have run voluntary educational savings accounts whereby employers and employees each contribute a proportion of their salary into a personal fund which can be used to support training and skills development (Folster 2001). The UK government introduced the Savings Gateway that was aimed at lower income households through the device of matching their savings with government contributions (McKay 2002).

1.2.3.2 Housing as Pension

A further dimension of the debates about asset-based welfare has included the role of homeownership, not least because it is an asset that is already widely distributed throughout the populations of advanced economies (see, for example, Groves et al. 2007; Regan and Paxton 2001; Ronald and Doling 2010; Sherraden 2003; Watson 2009). This has been quite explicitly taken up in proposals about the future of pensions. Thus, the World Bank, in re-affirming its support for a multi-pillar approach to pensions, has recently recognised that five pillars would appropriately reflect the range of national possibilities: a non-contributory 'zero' pillar; a contributory 'first' pillar related to income; a mandatory 'second' pillar in the form of an individual savings account; a 'third' pillar which could take a variety of forms including individual, employer-related, defined contribution, and defined benefit; and a 'fourth' pillar which took the form of intra-family and inter-generational smoothing of consumption in relation particularly to health care and housing (Holzmann and Hinz 2005).

The most recent of the EU Green Papers on pensions indicates that, in response to the challenge of maintaining pension commitments, member state governments have adopted a number of common policy objectives (European Commission 2010b). Central have been the raising of the age at which citizens are entitled to receive pension from the state, along with removing barriers to allow people to work longer, both combining therefore to encourage an extension of the working life and a reduction in the number of years claiming a pension, and so resolving the fiscal crisis by getting more tax receipts and spending less. Whereas the role of housing assets in contributing further to these solutions has not figured strongly in European-wide debates, the potential has not been entirely ignored. Indeed, over a decade ago there was some common understanding among representatives of the then member states that housing assets could provide a means of paying for people's needs in old age:

> In most EU Member States, older people live in owner-occupied housing. This means that many older people possess capital in the ownership of their homes. The Ministers were aware of the need to explore new ways of helping older people to safely utilize their capital (EU Housing Ministers 1999: para 9).

Such a statement begs many questions not only concerning how European households would view the use of their housing capital for the purposes specified but also how they would access it anyway. Here, in its recent Green Paper on pensions, the European Commission provides one answer in its suggestion that '(t)he Internal Market could also be helpful in extending access to additional sources of retirement income beyond pensions, such as reverse mortgages' (European Commission 2010b: 11). In other words, the way might be cleared for financial institutions to offer appropriate products, so that, while housing assets have not been at the very forefront of European policy development, they are on policy agenda.

1.3   Saving Through Housing: A Theoretical Framework

The approach in the DEMHOW project to understanding housing assets starts from the general question of why people invest at all? Why do people forego some consumption now, in order to save money for the future? There may be many different reasons why different individuals invest, including hoarding for the sake of hoarding, but following Keynes, Rowlingson et al. (1999) identify five principal motives:

1. To meet predictable future periods of low income such as retirement or high costs such as child rearing, referred to as the life cycle motive
2. To meet unpredictable future periods of low income such as unemployment or high costs such as a house fire, referred to as the precautionary motive
3. To leave an inheritance, for example, to children, referred to as the bequest motive
4. To make a future purchase of a particularly expensive item such as a car
5. To benefit from an investment opportunity to make particularly high rates of return

At least the first three of these motives can be founded in the life course. For biological as well as social reasons, the sustaining of human life continuously requires certain types of consumption: at the least food, water and warmth. In market economies, such consumption generally necessitates money payment. During certain periods of the life course, particularly in the early and later years, however, paid employment will not be a feasible option. Other periods without income are also possible, for example, as a result of illness or unemployment, and there will be yet other periods where, though income is being earned, individuals face large expenditures, for example, when raising young families. One of the general challenges facing the individual as well as the system as a whole, then, is the creation of surpluses at some periods in the life course that may be drawn upon during other periods of the life course: at some periods people may save in order that at other periods, even when they have no income, they can consume. What is sometimes referred to as consumption smoothing and sometimes as horizontal distribution over the life course are thus necessary conditions for long-term survival. Historically, human societies have developed a number of ways or mechanisms for achieving this.

1.3.1   The Life Cycle Model

The life cycle model (LCM), developed within the discipline of economics, can be thought of as a strategy for the temporal redistribution of income, the basic premise being that, at any one time, individuals set their present level of consumption in relation to their present level of wealth and their expected future income (Deaton 1992). In reality, because for most people the level of income over their life cycle approximates to an inverted U shape, the LCM can be characterised by a number of distinct stages. In the first stage, when individuals are young, their consumption (food, shelter, education and so on) generally exceeds their income so that they can be seen as, in effect, borrowing against future income. In the second stage, when they are in paid employment, their income exceeds their consumption, enabling them to save. In the final stage, that of retirement, consumption needs again generally exceed income so that individuals draw from their savings. The savings may take a variety of forms including cash deposits, antiques and equities, but may also include homeownership. So, generally, households have options not only about how much to save at any one point and how much in total they want to hold as well as about the balance between different forms of saving. The latter may be influenced by considerations of the expected risk and rates of return of different investment forms, as well as how liquid they are.

In most countries housing costs, whether renting or purchasing, constitute a major, perhaps the largest, single item in household budgets and therefore play a significant part in individual LCM strategies. Housing determines the standard of living of the household both directly through the house itself and indirectly through its impact on the amount of the budget left over to consume non-housing goods. In turn, because housing is expensive relative to incomes, it will also impact on the balance between consumption and saving. There is yet a further dimension in that housing is itself both an item of consumption and saving. Both renters and owners of dwellings each week consume a flow of housing services; this is their housing consumption. Owners additionally have an investment, which generally in advanced economies has experienced long-term growth and which is tradable. People who buy their own homes are therefore both consuming its services over time but also, provided that house prices are rising in real terms, building up their wealth, which may be drawn upon, or dissaved, later in life.

The economics literature has produced rich material for explaining individuals' portfolio decisions (e.g. McCarthy 2004). Nevertheless, empirical research acknowledges the gap between normative theory and social facts. Households do not necessarily make decisions as assumed in the standard suite of finance models, but may be influenced by, for example, herding behaviour, following the lead of others because they think they possess better information (Lusardi 2000; Campbell 2006).

1.3.2 The Welfare System

Any conformity between household behaviour and the LCM of course also takes place in a wider social, economic and political context, one aspect of which are the different society-wide arrangements, the national welfare systems, that also distribute savings and consumption. There are some goods and services that are essential to our well-being but are expensive relative to incomes, such as certain types of health care intervention, housing and education, and consequently require some way of smoothing payment over an extended period of time. There are others, also essential, but occurring at periods when as individuals we have no income: education and pensions, for example. For some individuals in some circumstances, it may be possible that payment can be met through their family, or financial institutions may advance a loan against a scheduled repayment programme. But, national welfare systems may be viewed as means of shifting the point at which something needs to be paid for from 'periods when they cannot pay to times when they can' (Glennerster 2003: 255). This is achieved by means of taxation that is paid into a single, central account which is drawn upon as required to meet specified expenditures on goods, services and transfers.

Though based on compulsory saving through the taxation system, the modern welfare state can then be seen as fulfilling some of the same functions as the LCM and financial institutions. The pattern of taxation, particularly based on earned income, achieves a distribution horizontally over the life span that effectively pays for many of the consumption needs of those with little or no earned income: the young who particularly consume education and health care and the old who particularly consume pensions, health, and social care. In reality, a large proportion of the welfare benefits received by individuals in many countries are self-financed, making each national welfare system a sort of 'savings bank' (Hills 1993: 19). The extent to which this occurs, however, will, depending on the nature of its welfare system, vary from country to country. It is well-established that there are large cross-national variations in levels of de-commodification, for example, and that these variations underlie distinctions between what are widely referred to as Liberal, Corporatist and Social Democratic regimes (Esping-Andersen 1990). The point here is that the links between welfare systems and the LCM can be expected to vary by regime type. It may be expected, for example, that in countries conforming to the Liberal type, where the state is viewed as a solution to welfare needs in cases where the market has failed, there will be greater responsibility on individuals to rely on an LCM-based strategy than those conforming to, say, a Social Democratic type, where the state is generally viewed as the first solution.

1.3.3 The Family

The institution of the family can also be seen as providing a means of horizontal distribution over the life course through an age, or life-stage, division of labour. Parents support their children by providing, out of their income, food, clothing and shelter, as well as assisting their emotional, social and educational developments. Grandparents may also contribute to care and socialisation of children during this pre-paid-work stage. As younger people age and move through the life course, they may begin to contribute to the acquisition of income as well as providing care and assistance to older family members. In life cycle model terms, then, people of working age, which roughly approximates to the childbearing age, invest for their

old age by having children. In welfare state terms, the family operates as a biologi-
cally-related, as opposed to a nation-related, welfare system in which their members
pool their resources of time, money and abilities to ensure the well-being of all not
only for life cycle savings but also precautionary and bequest motives.

The definition of the family and the ways in which it contributes to the support
of family members varies across countries (Attias-Donfut et al. 2005). In northern
and western Europe, the family is often equated with the household unit, generally
defined as those people who live in the same dwelling. Further, it is generally associated with the nuclear family, the ability of which to distribute horizontally across
the life course is more limited than the family as it is defined in many southern
European countries. Here, the family more commonly refers to a set of kinship
relationships that is both longer – the extended family, constituting all generations
living at any one time – and wider – incorporating siblings (Allen et al. 2004). There
are then important interconnections between the public dimension of horizontal
redistribution through the welfare system and the private dimension through the
family. The World Bank (1994) has argued that the former effectively 'crowds out'
the latter; in other words, the introduction of state provision of pensions reduces the
need for and willingness of adult children to support their parents. Albertini et al.
(2007) provide evidence that the patterns of intergenerational support are related not
just to the existence of state provision of pensions but also on the nature of that
provision which can be described by the welfare regime types.

1.3.4   Other Mechanisms

Societies have also developed a number of other ways or mechanisms for smoothing
consumption. These include civil society, the boundaries of which may be drawn to
incorporate individuals and groups within society beyond the family but not extending to the state or to the market. It may thus include friends and colleagues, as well
as trade unions, religious bodies, and formal third – or not for profit – sector organisations. In various ways, civil society may provide individuals with care and financial
support in periods of need as well as welfare goods and services, including education
and health care (Evers and Laville 2004). In assisting younger and older members of
populations, some of the activities of civil society may contribute directly to the horizontal distribution problem. In addition, and like welfare states, employers may also
operate forms of compulsory saving that smooth the consumption of their workers.
In effect, some of the remuneration package for the worker may be diverted into a
pension fund, health insurance, or education costs for the workers children.

1.3.5   The Mixed Economy of Saving

The term 'mixed economy of welfare' has been used by Johnson (1999) to refer to
the arrangements whereby any single welfare service may be provided by virtue of

more than one sector, by which is meant the state, the market, the family and so on. Taking pensions as a specific example, the World Bank's three-pillar model promotes one pillar based on government funding, a second on employment-based funding and a third on private saving (World Bank 1994). Here, the concept is adapted to refer not to welfare services but to the mechanisms – individual LCM behaviour, welfare systems and the family – through which households smooth consumption, or how they save and dissave. Households do not, therefore, necessarily rely on a single mechanism for smoothing consumption, but generally will make a selection of mechanisms with that selection being based, among other things, on perceptions of the advantages and disadvantages of each.

The mixed economy of saving can be identified at both the individual and the national levels. For both, it is apparent that the mechanisms are, to an extent, substitutable one for another. For example, in countries where the family provides a robust and effective means of ensuring the well-being of family members, there may have been less development of welfare state provision, while in countries with generous state pension provisions and social insurance protecting from the effects of unemployment, individuals may feel reduced need to accumulate private savings. The notion of the mixed economy of savings, applied at the level of the country provides a basis for classifying their approaches: for example, in some countries the main responsibility for smoothing consumption will lie with individuals and their ability to use their own resources to ensure their own well-being; in others, this will be achieved mainly by the welfare system; while in some others, the family will take a major role.

1.3.5.1 Financial Institutions

A further dimension of the mixed economy of saving lies in the role of financial institutions.

One of the functions of banks and other financial institutions is to borrow money from individuals and organisations wishing to invest money and to lend it to individuals and organisations that want to spend. They can thus contribute directly to the consumption smoothing strategies of households by providing loans to purchase commodities and services that may be particularly expensive in relation to income, such as houses or education, with repayment out of future income. In making available mortgage products, banks can facilitate the acquisition of homeownership and therefore of housing equity. In addition, they may provide a means of realising the asset tied up in housing, that is, dissaving, perhaps through a reverse mortgage or some other equity release product, and they may offer investment products such as private pensions that act as vehicles for households to save for their future, post-work, consumption. They therefore can contribute directly to the life cycle savings motive. Finally, financial institutions may offer insurance products that protect households from unanticipated costs, such as those arising from a large repair bill for their home and unanticipated loss of income, such as arising from unemployment; in that way, they offer a way of meeting the desire for precautionary savings.

1.3.6 The Role of Housing

How does housing, specifically homeownership, fit in this framework? This can be responded to in two steps. The first involves a brief review on how decisions about housing relate to the LCM. The second is the identification of the two income streams that may be derived from housing.

1.3.6.1 Housing and the Life Cycle Model

From an LCM perspective, it may be anticipated that households will invest in homeownership early in their working years both to secure somewhere to live (i.e. to consume housing services) and to invest for the future (i.e. to use homeownership as an investment vehicle). But, as they move through their life cycle, households will also be able to adjust their portfolios, which include moving up or down market in order to adapt the amount and proportion of their wealth they hold in the form of housing. In addition, subject to the availability of appropriate financial products, they may be able to reduce their housing wealth by withdrawing some of the equity of their home.

Changes throughout the life cycle in the acquisition of housing and non-housing assets and debts, including mortgages, may be seen as a result of the strategy of (more or less) forward-looking households taking into account the possible impact and fluctuation of public policies. For example, changes to interest rates may be an incentive for households to modify their portfolio structure. Also central to our theoretical framework, however, are the links with welfare systems and family relationships. In relation to the former, it can be expected that in general the propensity to invest in housing assets, as opposed to consuming or investing in non-housing assets, is linked to the support expected in old age from the state in the form of pensions and other provisions. But there will also be a link to the housing system in each country, specifically to the availability and affordability of different forms of housing tenure, which, as will be shown in Chap. 2, is roughly correlated with welfare regime type. In relation to the latter, it might be expected in turn that investment in housing assets will tend to be higher in those countries where the extended family model continues to be significant and where homeownership is often deemed to be important to the wider family project in providing both a physical and an emotional space for the family (Allen et al. 2004).

1.3.6.2 Income Derived from Homeownership

Whereas housing can be viewed as a store of wealth, the way in which that wealth is convertible into income, and therefore consumption, is also important. In this respect, those who own their homes outright, which, given the nature of many mortgage products, will often be people later in their life cycle, have two, non-mutually exclusive, possibilities: an income in kind and an income in cash.

An income in kind, often referred to as imputed rent, represents the flow of services derived from the size and quality of the dwelling and its location. Dwellings that are larger, have more expensive and desirable facilities, and are in locations with good access to desirable land uses will tend to attract higher market prices than small, poorly equipped dwellings in unattractive locations. This market price can, in turn, be thought of as proportional to the rent that the dwelling would attract. Owner occupiers, while they do not literally pay rent to themselves, nevertheless may be considered to receive a flow of services with a value equivalent to the rent that the dwelling would attract were it on offer in the market. In that sense, they receive an income in kind.

An income in cash lies in the capital value or equity of the dwelling. Insofar as dwellings are tradable commodities, each will have a market price that represents its valuation as a capital asset. People who acquire the ownership of a dwelling thus have a store of wealth, the value of which may increase or decrease over time, and, like other stores of wealth, provide opportunities to realise the value in the form of cash. These opportunities cannot always be easily taken up. Nevertheless, they might be achievable by selling the house and moving to a cheaper house, that is, releasing some of the equity, or moving into a rented property, thereby releasing all the equity. An alternative is to remain in the home while using a financial product such as a reverse mortgage or equity release loan that will provide a cash income.

1.3.7 Cross-Country Variations

Central to this schema is a notion, therefore, that the way in which households, in aggregate, build up housing assets and the extent to which they use them in old age is influenced by national characteristics, what can be termed the formal institutions: the supply of housing finance products, the scale and nature of welfare provision for older people and so on. It therefore follows that from the patterns of formal institutions in each country it may be possible to understand average household behaviour in each country.

1.4 Methodologies for Researching the Three Questions

Starting from the perspective of a framework based on the notion of a mixed economy of saving, then, how has the research task been approached? Broadly, our three questions have been addressed from two different ontological and epistemological positions, each using different methods of data collection. This question-methods fit, combining quantitative and qualitative analyses, constitutes a 'mixed-methods' approach (Creswell and Plano Clark 2007), the rationale for which, following Smelser (2003), can be that the best methodological strategy in comparative research is to get a foothold wherever possible and thus rely on multiple data and methods to provide answers.

The utilisation of this framework is nevertheless limited by a number of constraints and choices. Fundamental is that the research cannot both be restricted to just a few member states and, at the same time, produce results that are representative of the full variation across all of them. This presents a number of practical challenges of which foremost are data availability and cost. The first of these has already been alluded to in earlier parts of the present chapter discussing trends in relation to homeownership. Notwithstanding the work of Eurostat and other international statistical agencies, many variables that would be appropriate for statistical analysis are not available. This may be because they are not centrally collated, or they are centrally collated but the definitions used are peculiar to each country – that is, they are not harmonised – or they are available and harmonised, but only for the larger countries (usually some of the older member states), or they are available for a few years only, thereby not allowing the identification of long-term trends (usually the case in the newer member states). In reality, therefore, the statistics base for establishing relationships is often possible on the basis of only a few, generally the larger, and western, member states. In these circumstances, generalising across all member states demands considerable caution. Alternatively, measures for all member states may be possible but based on very broad approximations so that caution in interpretation is required in this also.

The second limitation refers to the resources required in order to collect primary data in and about all member states. This is not simply a problem that if X amount of resources are used to study one country, each time another country is added to the list of those studied there is a further X amount of resources required. Rather, each time another country is added, the problems of harmonising across language and culture multiply. In any case, referring to household level data, any sample size that would allow generalising to all the households in even one country would entail an amount of resources which exceeded the research funding available. Effectively, this rules out large sample surveys or interviews. Overall, the data problems might be summarised as too few variables, too few countries and too few households.

Question 1
The past behaviour question: how have Europeans used housing assets in the past? The objective is to establish the extent to which households in different member states have actually accumulated and used housing assets, and especially how that use has extended to contributing to income needs in old age.

This has been addressed principally by accessing, presenting and analysing secondary sources in the form of macro and micro data sets, the principal ones being listed in Table 1.3. The OECD data cover only the OECD countries. Frequently, in our analyses, we have been able to use only a very limited number of these countries, simply because harmonised data have not been available even for all OECD countries. For the analyses at the micro level two European datasets, EU-SILC and SHARE were used. The EU-SILC data set covers the EU27 and contains information on households and housing though containing little on all assets in their portfolios. Data on health and also on the composition of the asset portfolio, however, are available in SHARE, but unfortunately this is limited to households aged

Table 1.3 Characteristics of main data sets

	OECD	EU-SILC	SHARE
Level	Macro	Micro	Micro, only households of 55+
Data	Economic indicators, assets	Income and living conditions	Income, assets, household structures
Countries	OECD, diverse selections	EU27	Nine European countries

over 55 years and to a small number of EU countries. Together with existing literature, mainly itself comparative, these sources have enabled a picture to be built up of the main trends and relationships affecting the building up and using of housing assets across Europe.

Question 2

The present attitudes question: what are their current views and future expectations about how they should be used? Against the background of revealed behaviour, how do European households view housing equity in their life strategies and what are the similarities and differences between countries and generations?

Understanding household attitudes and strategies towards housing equity and drawing conclusions about country similarities and differences requires acknowledgement of the different institutional contexts. This raises the question of what is context and how can it be researched? In adopting the institutional approach, as described by Kato (1996), we have assumed that household behaviour is bounded by both formal and informal institutions (Toussaint 2011; Kato 1996; North 1991). By formal institutions are meant laws and rules that are included in national legislation and policies such as tax and pension policies, while informal institutions are the unwritten rules of societies, their values and norms, that are determined outside the formal channels. They can be self-reinforcing through mechanisms as imitations and traditions; and they also serve as sanctions (example of those sanctions are measures that have consequences for community membership and reputation) that facilitate the process of self-reinforcement (Tridico 2004). For example, the question of why older people do not dissave housing equity, may be attributed to both tax policies as well as values about having a debt.

Knowledge of the formal and informal institutions in each country has been built up from existing published sources including scientific literature and policy documents. This has provided a background for interviews undertaken with samples of households. These have enabled the building up of an understanding of the role of formal institutions such as pension and tax policies and also the role of informal institutions (Elsinga and Mandic 2010). Whereas this has been based on following a list of questions about thoughts, reasoning and experiences, some were also based on vignettes, that are descriptions of situations of which interviewees are asked what they would advise the persons in the situations to do, and why. This approach enables the interviewees to feel free from the need to justify their own situation.

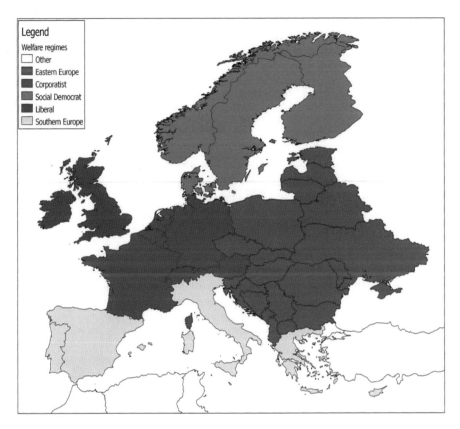

Fig. 1.2 Different types of welfare regimes

1.4.1   Selection of Cases

Given that context matters, and that undertaking household interviews in all 27 member states would be prohibitively expensive, and because welfare regimes for care and income in old age are central parts of that context, a set of eight countries was selected as representative of the main types of welfare and housing systems. This selection was based on the not uncommon extension of the Esping-Andersen typology (Esping-Andersen 1990) to a total of five regime types (see Fig. 1.2). It included a Liberal welfare regime (UK), two Corporatist welfare regimes (Germany and Belgium), a Social Democratic welfare regime (Finland), one mix of a Social Democratic and Corporatist welfare regime (the Netherlands), a Mediterranean regime (Portugal) and two former Eastern European countries (Slovenia and Hungary).

This selection also covered a range of countries with different homeownership rates. For example, in Hungary and Slovenia, most people in all income groups are homeowners. In the Social Democratic and Corporatist countries selected, by contrast, the majority of the highest income households are homeowners while the majority of lower income households are tenants.

Whereas this was a device for reducing the number of countries in which interviews were undertaken while not losing the broad range of national situations and circumstances, it was also a more general device for capturing the EU-wide picture. Accordingly, throughout the book, results are, wherever possible, presented in these five clusters for as many countries as possible. This means some figures will contain maybe eight countries and others 27.

In each of the eight countries, a sample of about 30 households was selected from an area of average economic development. In the selection of households, age and children played a key role since we expected different generations to have different views, and also that people with children have different ideas from people without children. Accordingly, the selection consisted of 10 households in their 30s, 10 in their 50s and 10 in their 70s, each group having been made up of 6 with and 4 without children.

Question 3

The policy outcome question: how would housing assets perform as pensions? In other words, if housing assets were systematically used in order to meet income needs in old age, how would it measure up to standard criteria against which pension systems are generally assessed?

Our third research question has been approached in three ways. Firstly by, the quantitative and qualitative data collected for Questions 1 and 2 provide an initial understanding of how housing assets have been used in old age. Secondly, the published statistical information has enabled an analysis of what housing assets held by older European households could, if systematically used, contribute to their income. This contribution is evaluated against a set of criteria commonly used to measure the strength of pension systems. Thirdly, an analysis of Asian experiences is undertaken. Japan, Singapore and Korea – countries with a level of economic development that is comparable to Europe – all have experience in acknowledging a role for housing equity in their welfare and pension policies. In that sense, they have been forerunners of property asset based welfare systems, thereby offering the potential for learning policy lessons.

Table 1.4 provides an overview of the different approaches and the data used in the project. Through the use of quantitative research at the macro and the micro level, a picture of the statistical relations is built up. By adding the qualitative picture we are also able to answer questions on how and why things are the way they are. Together these different approaches provide answers from different angles to the question what is the role of housing equity in an ageing Europe.

1.4.1.1   Economic and Financial Crisis

At this point, a final note about methodology is appropriate. In dealing with housing as a pension, the situation in housing markets and financial markets is of great relevance. Our examinations of past behaviour and, to an extent, even present attitudes are

Table 1.4 Overview of questions and data

Questions	Micro	Macro
Question 1: How much housing wealth accumulated and released?	EU-SILC, 2008; SHARE, 2005–2007	OECD, diverse years and selections
Question 2: What are households' attitudes and strategies and how do countries compare?	In-depth interviews in eight countries, 2009	Institutional studies in eight countries, 2009
Question 3: How would housing perform as a pension?	EU-SILC, 2008; SHARE, 2005–2007	OECD, diverse years and selections
	In-depth interviews in eight countries, 2009	Institutional studies in eight countries, 2009
		Studies in three East Asian countries, 2009

based on housing and financial markets during the long period of growth in the 1990s and 2000s. This has changed dramatically. The economic and financial crisis from 2007 has resulted in house prices in many areas coming under pressure or even decreasing. Thus, there is less housing wealth than there was. Moreover, the behaviour of lenders in most countries is different with them being more careful in providing mortgages, while households may have changed their strategies and expectations towards housing assets.

In respect of this the 2010 EU Green Paper concluded that:

> The financial and economic crisis has seriously aggravated the underlying ageing challenge. By demonstrating the interdependence of the various schemes and revealing weaknesses in some scheme designs it has acted as a wake-up call for all pensions, whether PAYG or funded: higher unemployment, lower growth, higher national debt levels and financial market volatility have made it harder for all systems to deliver on pension promises (European Commission 2010a, b: 6).

Unfortunately, this shift affects the DEMHOW project. Most of the data used for the quantitative analyses were collected before the crisis and cover a period of economic growth and increasing house prices. The description of the institutional contexts and in-depth interviews took place in the summer of 2009 in the middle of the financial crisis. This material has allowed the drawing of some conclusions on the effect of the crisis in the financial markets as well as some observations about what may be the impact of the crises for housing as a possible pillar in pensions. It is nevertheless a cause for further caution in interpreting the research findings.

1.5   Content and Structure of the Book

Chapters 2, 3 and 4 provide responses to our first and second questions. Each draws both on analyses, using macro and micro level data, of observed behaviour in the past, as well as on context-based interpretations of qualitative interviews about past, present and future attitudes and intentions.

Chapter 2 draws on a number of bodies of literature in order to provide insights into the two trends in national homeownership rates: the tendency over recent decades for these to increase across Europe alongside continuing national differences. The political science–housing studies literature has stressed the importance of a trade-off between homeownership and state welfare provision, especially in relation to pensions. Essentially, countries with high homeownership rates tend to have smaller state pension commitments. The economics and housing studies literature suggests the importance of other factors. At a macro level, the growth of housing finance markets, itself generally an outcome of de-regulation has facilitated higher levels of access to the tenure. National housing policies too have made a difference and explain why homeownership rates differ so much from one European country to another. Set against this, the evidence from our interviews both supports the importance of the formal, national institutional framework within which individual households operate, and also suggest the importance of informal institutional frameworks.

Chapter 3 presents evidence showing that housing equity constitutes the largest single form of wealth for the average European household, particularly so for older Europeans. In comparison with the ownership of shares, housing equity is both considerably larger and more evenly spread. Limitations in the availability of suitable data do not allow strong conclusions about explanations for national differences in the role of housing in the composition of household wealth. However, across a very small sample of countries larger welfare spending on older people is associated with lower levels of non-housing assets in the household portfolio, suggesting the possibility that the relationships, discussed in Chap. 2, about a trade-off between welfare expenditure and homeownership, may need to be qualified: the response to a lack of generosity in state welfare spending may not necessarily be the acquisition of more housing wealth, but of more wealth in forms that may be more easily realisable.

The investigation of the role of mortgage debt has been similarly restricted by data availability considerations. Nevertheless, it is clear that the reliance on mortgages varies from country to country. In part, this is related to regime type, with a high level of mortgages in Liberal and Social Democratic countries and a low level in the Mediterranean and Eastern countries. But it also seems related to cultural perspectives on debt: in many countries, for example, debt is something to get rid of as soon as possible rather than a portfolio decision.

Chapter 4 shows that European households appear to have reduced their total wealth progressively through their retirement years, in effect enhancing their pensions and in that way acting consistently with the LCM. The dissaving of housing assets, in contrast most categories of wealth in individual portfolios, however, is complicated because housing is both a consumption and an investment good. Indeed, examination of a number of statistical sources together with the evidence of previous empirical research shows that the most common adopted strategy is not to dissave housing assets at all.

One reason for this appears to be that housing is often viewed as a substitute not only for the perceived inadequacy of government pension provision but for government provision in other areas, especially social and health care needs. For many, the home is also seen as a bequest, to be left to children, and many have the notion of a debt-free, ownership ideal.

Chapters 5 and 6 bring together responses for the third research question about how housing assets might perform as a pension. Chapter 5 focuses empirically on Europe and considers, using published data, the extent to which the use of housing equity to provide both an income in kind (imputed rent) and an income in cash (through equity release) might contribute to pension needs. The estimates presented in this chapter indicate that housing income, both in kind and in cash, do and could make a considerable contribution to maintaining the former standard of living of older people. Income in kind alone appears to constitute, on average across member states and households, about a quarter of total income. It is clear that if to this were added an income in cash created by realising the full equity of the home, the boost to existing incomes would indeed be considerable. It follows that, all other things being equal, the use of housing in a way consistent with the LCM, would result in the average older European enjoying a higher standard of living.

But here there are also limits to the benefits brought from using housing assets in this way. There appears to be a positive correlation between those with most housing assets and those with most cash income anyway. In other words the potential of housing equity would be greatest for those who, on objective grounds, might be deemed to need it least. Generally, housing assets would not appear to be a mechanism for reducing inequality across populations; they may help to smooth income across the life cycle, but not across income groups; they constitute a means of horizontal not vertical redistribution. Moreover, given doubts about the future sustainability of homeownership sectors at their present size, it is possible that the number of non-homeowners may actually grow.

Chapter 6 examines what lessons might be learned from the experience of the economically more advanced countries of East Asia, the rationale lying in the intensity and longevity of homeownership policy pathways: East Asia can be seen as a policy pathfinder, providing Europe with an opportunity to learn lessons from those who have gone before. The broad message is that homeownership can work relatively well as a pension in a context in which the wider welfare system is not based on a strong commitment to redistribution. In such contexts pension systems and ownership mutually reinforce.

But the East Asian experience also demonstrates limits to the reliance on housing. Homeownership may facilitate distribution across the life cycle and even across generations, but it is not generally a vehicle for distribution across income classes: the tendency in recent decades is for housing markets and property ownership to reinforce social inequalities rather than alleviate them. In general, those with least housing assets will have least non-housing income and wealth. If housing fails on the adequacy test, it may also fail on the sustainability test. It has become clear that pre-1997 systems were built on the assumption that house price increases could outpace inflation in perpetuity. The reality of house price deflation results in part in additional pressure being placed on government to protect housing markets; house prices become an intensely political issue. In other words, governments cannot simply deflect responsibility for the well-being of older citizens by promoting homeownership, if that particular form of investment does not perform in a way that assures well-being.

Chapter 7 brings together the findings from the earlier chapters. These are broadly consistent with the mixed economy of saving, the interplay of the LCM, the family and the welfare systems that provide households with different ways of achieving horizontal distribution across the life cycle. These possibilities along with the influence of informal institutions can be seen through the distinctive patterns of behaviour in each of the five welfare regimes. For example, in the Eastern regime group where homeownership has grown to very high rates and welfare spending generally is low, often their populations do not consider their pension systems to be safe. Generally, housing equity has become the ultimate precautionary fund. The family plays an important role with adult children often providing support to their parents with the understanding that the family home will become theirs: in that way, housing equity is part of the family strategy. In contrast, the Social Democratic countries have relatively generous welfare systems offering considerable protection for their citizens. Housing equity has not played such an important role: there are fewer homeowners and homeowners build less housing equity, with many people extending their housing loans into their retirement years.

Alongside these country differences there are common variations across generations. The pre-baby boom generations often appear very cautious, being reluctant to spend their assets on consumption and eager to pass on an inheritance to their children. In contrast, baby boomers and later generations commonly appear much more willing, sometimes eager, to continue, if not increase, the level of consumption they had enjoyed while working, and if this could be achieved by using the equity in their home that was acceptable. Younger age groups are often even more open to the necessity to have to find their own solutions to their income needs in older age and to use housing assets to do so.

These geographical and generational outcomes are important in setting a context in which the policy implications of developing a greater reliance on housing assets in providing a contribution to income in old age can be established.

Chapter 2
Homeownership Rates

2.1 Introduction

Across European countries, homeownership has demonstrated both convergence and divergence. In recent decades, homeownership rates have increased. Having numerically superseded all forms of renting, it is now the majority tenure providing the living circumstances for about two-thirds of European households (Doling and Ford 2007). Whereas this may seem to point to a convergence in housing systems, there remains evidence of considerable divergence. The growth over time has not occurred equally over countries. Europe's largest economy, Germany, has the lowest rate of homeownership, considerably below the point at which it would be the majority tenure. In contrast, a number of other countries, notably Hungary, have homeownership rates in excess of 80%.

The overall aim of this chapter is to present the developments in homeownership sectors in rather more detail and then to provide explanations of the common growth and different outcomes. In this, it responds to aspects of both the past behaviour and present attitude questions. The chapter begins, then, with statistical information about tenure trends by EU member states over time. The main part of the chapter considers explanations underlying the popularity of homeownership.

In this, we build on several bodies of research. The first is grounded in housing studies and political science, and specifically in the seminal work of Jim Kemeny and Francis Castles. Both explored the relationship between homeownership rates and welfare provision, based on the notion of a trade-off such that countries with more homeownership are those with the lowest levels of social protection, with the one seeming to substitute for the other. This body of work thus fits with a basic proposition underlying the present book that homeownership and welfare states are both forms of horizontal distribution, laying down investment during working years that enable consumption during retirement years.

Another body of research has also been present in housing studies literature but draws much of its inspiration from the discipline of economics. This has identified

J. Doling and M. Elsinga, *Demographic Change and Housing Wealth: Homeowners, Pensions and Asset-based Welfare in Europe*, DOI 10.1007/978-94-007-4384-7_2, © Springer Science+Business Media Dordrecht 2013

a number of drivers or factors that appear to have influenced both the growth of homeownership rates over time as well as the continued country differences. At a macro level, the main factors identified are those relating to developments in financial markets which have broadened access to housing loans, thus further enabling consumption smoothing, and the impacts of national housing policies, which have altered the balance of advantages and disadvantages accruing to different tenure options. The importance of at least some of these factors, especially the relative opportunities and costs of different tenures, is supported by the results of household interviews carried out as part of the DEMHOW project. At the micro level, the main factors relate to household characteristics, such as age, income and marital status, all contributing to understanding of responses to the tenure options.

2.2 Homeownership Across Countries and Time

The exact enumeration of homeownership rates in each European country, and hence comparison between countries is not possible, largely due to the absence of systematically collected and fully harmonized data across the countries over time. Definitions of what constitutes homeownership differ from country to country, for example, there are debates as to whether to include Swedish cooperative housing as a form of homeownership (Stephens 2003). However, even tolerating an element of imprecision, it is clear both that homeownership has grown across the EU over time, and that homeownership is now the predominant tenure in the EU. In 1945, homeownership was a minority tenure in each of, what are now, the 27 countries of the European Union. By 2003, homeownership was the majority tenure in every country, except Germany (Table 2.1).

It is also clear that there are very large variations across countries at any one point in time. The data in the final column of Table 2.1 show that in some countries, Bulgaria, Estonia, Hungary, Lithuania and Romania – all part of the former communist bloc – homeownership is almost universal. In a number of other countries, including at least some of the older member states, the rate is above 80%, whereas some of the northern and western countries, including the Netherlands and Denmark have rates below 60%. Clearly, then, each country in the EU has a different pattern of housing tenure and with that different opportunities for households to become homeowners. The challenge taken up in the rest of this chapter is the presentation of explanations for those variations.

2.3 Homeownership Rates and Welfare: A Trade-Off?

2.3.1 Homeownership and Social Spending

The argument that national rates of homeownership could be explained by reference to welfare systems was developed by Jim Kemeny (1980). He argued that western countries could be located along a privatism (individual responsibility)-collectivism

Table 2.1 The post-war growth of homeownership: percentage share of total stock by year

	1945/1950	c1960	c1970	c1980	c1990	c1995	c2002	2006–2009
Corporatist								
France	–	41	45	51	54	54	55	57
Germany	–	–	–	–	38	38	42	–
Austria	36	38	41	48	55	41	56	56
Belgium	39	50	55	59	67	62	71	78
Netherlands	28	29	35	42	44	47	53	57
Social democratic								
Denmark	–	43	49	52	51	50	51	54
Finland	–	57	59	61	67	72	58	59
Sweden	38	36	35	41	42	43	42	66
Mediterranean								
Greece	–	–	–	70	77	70	83	80
Italy	40	45	50	59	67	67	80	80
Spain	–	–	64	73	76	76	85	85
Portugal	–	–	–	57	58	65	64	76
Liberal								
UK	29	42	49	56	68	66	69	70
Ireland	–	–	71	76	81	80	77	75
Eastern								
Estonia					35	37	95	98
Latvia					22	39	82	87
Lithuania	–	–	–	–	47	87	95	97
Bulgaria					93	93	93	97
Czech Republic	–	–	–	–	59	62	64	–
Hungary	–	–	–	–	78	89	92	92
Poland					51	56	55	75
Romania					76	89	95	96
Slovakia					73	74	89	88
Slovenia	–	–	–	–	68	88	82	82

Note: dates are approximate
Sources: Catte et al. (2004), Scanlon and Whitehead (2007), EMF (2010), MRI (1996), Balchin (1996)

(state responsibility) spectrum and that their position on it influenced the nature and extent of their government's policy orientation towards private forms of tenure (homeownership) as in the case of Australia, collective forms (social renting) as in the case of Sweden, or mixed private and collective as in the case of the UK. Schmidt (1989) provided a statistical description and test of Kemeny, finding a significant correlation across western countries indicating that those with higher proportions of their public expenditure devoted to social protection measures such as sickness and unemployment benefits had smaller home owning sectors. This finding has been more recently confirmed by Conley and Gifford (2006) using a data set that included not only many western countries, but also some former communist countries, specifically Hungary and Poland.

Table 2.2 Homeownership rates by social protection expenditure on older people

		Homeownership (1980s data)	
		Low	High
Social protection expenditure on the aged (1990)	Low	Japan Portugal	Australia Canada Finland Ireland New Zealand Norway Spain USA
	High	Austria Belgium Denmark Germany Luxembourg Netherlands Sweden	Greece Italy UK France

Source: Castles and Ferrera (1996)

A more specific relationship between welfare spending and homeownership rates was proposed by Frank Castles, coining the phrase 'the really big trade-off' to describe the relationship with state expenditure on pensions. Prima facie evidence of this is provided in Table 2.2, where, the cut-off point between high and low rates of homeownership is the median value, which in this period was 63%, and the cut-off point between high and low expenditure on pensions as a proportion of GDP is also set at the median value, which was 8%. The fact that most countries are clustered in the top right-hand and bottom left-hand cells is consistent with the notion of a homeownership–public pensions trade-off. Essentially, many of those countries that have developed high homeownership sectors have also experienced relatively low rates of state spending on older people, and in that sense there appears to be some mechanism, or mechanisms, whereby less of one is compensated for by more of the other.

The housing-pensions trade-off suggests a degree of functional equivalence between the two elements in that they both necessitate investment during the working years of the life cycle, that is, out of income earned through work setting money aside in order to store up wealth that may be drawn upon in the retirement years. While the means by which these investments are secured may be different – home-ownership generally being an individual and voluntary investment decision with state pensions being a form of forced saving through taxation systems – they can be seen as offering substitutable outcomes: 'the private ownership of housing and the public provision of aged pensions constitute alternative means of horizontal, life-cycle redistribution by which individuals guarantee their security in old age' (Castles and Ferrera 1996:164).

The notion of a trade-off could, of course, also be extended to other vehicles for horizontal distribution across the life cycle, such as investments in personal (as opposed to social) pension funds, and in stocks and shares. Arguably a more contentious issue concerns the processes by which such trade-offs are effected. In presenting his findings, Schmidt argued that the correlation should not be confused with causation, and that it is not clear which is cause, which is effect, or whether both are the product of some third factor. This uncertainty has generated some speculation about what the processes might be, and following Fahey (2003), it is possible to identify from the literature two main propositions.

The *constraint* induced trade-off comes about because, as Kemeny (2001) argues, 'how housing is paid for varies greatly between different forms of tenure' (62) with tenants generally being able to spread the costs over the life course, 'whereas owner occupation concentrates them in the early stages of the life course' (62). But, the front loading of house purchase falls heavily on young families often when their incomes have not developed to their fullest potential and when they anyway face the additional costs involved in child rearing. The argument is that in these circumstances something has to give:

> …house purchase and the social insurance contributions that fund pensions are simultaneously the two biggest items of expenditure that confront families across the life-cycle. Hence the trade-off is not just theoretical, but actual; other things being equal, the more taxes one pays for a high pension in old age, the less one can afford for housing purchase and *vice versa* (Castles and Ferrera 1996:164).

The *need* induced trade-off occurs because 'income streams available to the old in some countries by virtue of social security entitlements may in other countries be available by virtue of private savings, private insurance, or through equivalent benefits stemming from property ownership' (Castles 1998b: 205). In other words, when people, because they own their home, can live rent free, they can make do with smaller pensions. Following this logic, a number of studies have shown that once income in kind from housing is added to income in cash from pensions, the total income of homeowners is higher than from income in cash alone and the unequal distribution of income among older people may be considerably modified (e.g. Castles 1998a; Ritakallio 2003). There is sound empirical support here for the proposition that homeownership provides the owner with an income, albeit in-kind income, that can be considered to be a sort of pension.

2.3.2 Homeownership and Welfare Regimes

Notwithstanding the empirical evidence supporting a negative relationship between homeownership and social spending, the connection with our five regimes is less clear. The empirical basis of Esping Andersen's three worlds was partial, focusing on pensions and excluding housing, but in practice his typology has been widely used as a means of organising understandings of the full spectrum of welfare goods and services. In attempting to map housing onto the three worlds, Hoekstra (2003), however, concluded that his typology cannot be simply applied to housing. Moreover, he argues that even when the typology was extended, by adding Mediterranean and Eastern groupings of countries, the housing-regime connections were not marked.

Much the same conclusion comes from testing tenure expectations derived from Kemeny's work. From this, it would be expected that the Liberal countries would have high homeownership rates, and Table 2.1 indeed shows that UK and Ireland have rates above the European average. In contrast, while it would be expected that the Social Democratic countries would have low homeownership rates, this holds for Denmark and Sweden, but not for Finland prior to the last decade or to Norway, which, although not included in the table, has a homeownership rate of 77%. In the Corporatist countries, with their focus on rental provision, a modest homeownership rate could be expected, which holds for Germany and Austria but not for Belgium.

Likewise, the scatter plot in Fig. 2.1 indicates this partial degree of group homogeneity. In 1995, the Social Democratic grouping is neatly clustered as are, with the exception of Belgium, the Corporatist countries. The remaining countries, all from western Europe, are jumbled in the top left corner of the graph, indicating overlap between the Southern and the Liberal countries. By 2007, there are somewhat less clear distinctions. This now also includes the Eastern countries – Slovenia, Hungary and Poland – which are located among the Liberal and the Mediterranean countries. Although overall the relation between social expenditure and homeownership still exists ($r=0.30$), it is weaker than in 1995 ($r=0.44$) and the Corporatist and Social Democratic clusters disappeared. The relation between homeownership and social expenditure, and the link to welfare regime types, appears to have shifted. Probably the most that can be justifiably argued from these data, then, are that countries in the Corporatist and Social Democratic groupings tend to have low homeownership rates and high social spending, with countries in the other regime groupings having high homeownership rates and low social spending.

2.4   The Drivers of the Homeownership Decision

Whereas the recognition of a trade-off between homeownership and social spending provides an understanding of one macro relationship, it contributes little about other aspects of national contexts within which households make decision about tenure, nor indeed about how different individual and household characteristics might figure. Here, there are several bodies of literature – from, for example, economics, housing studies, demography – that collectively suggest the importance of a number of drivers or influences. At the macro level the nature and size of housing finance markets and housing policy regimes have been changing and help to explain the growth of homeownership rates. At the same time, they form parts of national contexts so that they also inform an understanding of the choices of households in these countries and the national differences in outcomes. At the micro level, drivers, such as income and age, relate to the characteristics of individuals, which may constrain or open up opportunities.

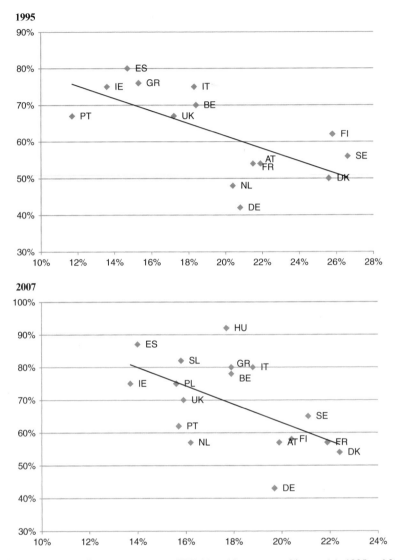

Fig. 2.1 Social expenditure as percentage GDP (x) and homeownership rate (y), 1995 and 2007

2.4.1 Housing Finance

From the 1980s, housing finance systems in the EU have been going through radical reforms, which have included new methods of funding mechanisms such as securitization and the introduction of innovative mortgage products.

The trends affecting mortgage markets have also been similar across countries. Mortgage markets have been liberalized in many western European countries over the last 20 years as part of the more general globalization of finance markets: restrictions on the use and terms of loans have been lessened, and a wider range of financial institutions is now permitted to offer mortgages. An important goal of deregulation was to improve the efficiency of the system by opening up the market to new providers and increasing competition amongst lenders, thereby lowering costs to consumers (Scanlon et al. 2008:110).

The general trend towards the deregulation of the housing finance system has made mortgage products more affordable, enabling more people to consume – and invest – in advance of saving, and thus it has generated further spread of homeownership (Apgar and Xiao Di 2006).

Figure 2.2 provides a representation of the extent and growth of mortgage lending in relation to GDP. There are systematic differences according to welfare regime type. Overall, mortgage markets remain quite underdeveloped, with ratios generally below 20%, in the newer member states, and comparatively underdeveloped, generally between 20% and 40%, in the corporatist regime countries. In general, mortgage markets are much more developed, however, in both the Social Democratic and the Liberal countries.

Across the entire EU, the average mortgage stock to GDP ratio increased from 31.9% to 48.2% between 1998 and 2008. The country patterns differ: at one extreme, in Germany, the outstanding mortgage level decreased slightly, while in the Netherlands, it increased by 38% points. In the new member states, the mortgage to GDP ratio is lower than the EU average (in Hungary it is 14% and in Slovenia 9.1%), but the increase was very fast over the last decade.

These shifts in housing finance systems occurred in the funding schemes of housing loans, and in the loan products and loan contract enforcement. These changes were a result not only of measures created and implemented by public policies, but also of the innovation of the banking sector.

2.4.1.1 Funding of Mortgage Loans

The dominant funding source in Europe is deposits that households put into banks as savings or current accounts. One estimate is that about two-thirds of the mortgages in Europe are funded by such retail deposits (EU 2006). The second largest funding source is the covered bond, a debt instrument that is secured by mortgage loans as collateral to which investors have a preferential claim in the event of default. Covered bonds represent about 15–20% of mortgage funding in Europe. The third largest source is mortgage-backed securities, which have been more popular in the USA, but in Europe account for only about 5% of mortgage funds (EU 2006).

Since the end of the 1990s the rate of increase of mortgages has been much higher than the increase in households' deposits, creating a funding gap. In the Euro area, between 1999 and 2006, the total deposits increased by 33%, and outstanding mortgage loans by 83%, which means that the share taken by covered bonds and MBS has increased. The European Central Bank pointed out, however,

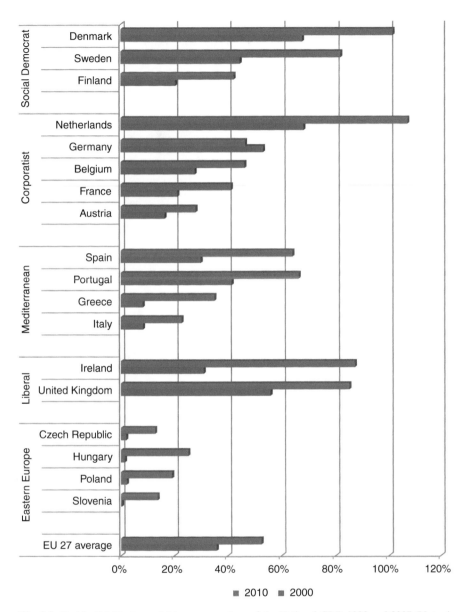

Fig. 2.2 Residential Mortgage debt as a percentage of the National GDP, 1998 and 2007 (Note: *
Finland, Austria, Czech Republic, and Hungary: 1998 = around 2001–2002) (Source: Hypostat 2010)

it would be misleading to try to establish unidirectional causality here, running from an
increased funding gap to a diversification of the funding sources. Indeed, part of the growing
funding gap is actually explained by the existence of the alternative sources of finance,
which allowed banks to expand their loan market against backdrop of increasing demand
and higher competitions (ECB 2009:41).

The expansion of the mortgage market, then, was fuelled by the availability of relatively cheap resources on the world market, which sought investment possibilities through the bank system. The extent to which this occurred, however, varied from country to country. Legally, only in Belgium is it not possible to issue covered bonds, while in Slovenia, it is legally possible, but, in practice, not used. Traditionally, Germany has had Pfandbrief since 1927, and it has 34% of the entire European market of covered bonds (ECB 2009).

2.4.1.2 The Innovation in Loan Products

Innovations in loan products have also contributed to the expansion of the mortgage market and the growth of homeownership. While each country moved to a more liberalized system, there is a large variation in what their systems offer. Overall, however, the innovations have increased the availability of mortgage loans, including changes in the type of the interest rate (such as variable rate mortgage and interest-only mortgages) and modifications of repayment structure and terms of the loan (such as introducing 30 years loans). A consequence of these changes has been to make loans affordable for a wider range of households including those with low incomes who could not afford owner occupied housing earlier. From the end of the 1970s to the early 1990s, homeownership rates for younger age cohorts increased most in those countries in which the deposit requirement was least (Chiuri and Jappelli 2006).

2.4.2   The Relative Attractions of Home Owning and Renting

Developments in the availability of housing finance are part of a wider context in which the balance of advantages and disadvantages of different tenures is established. For individual households, housing opportunities will not only be influenced by the availability and cost of finance but also the nature and effect of an array of market conditions and policy effects.

2.4.2.1 Tax Policy and Other Subsidies for Homeownership

At least in the short run, interest rate subsidies can be expected to increase the affordability of loans and the level of mortgage debts taken up. There appears to be a great variety in the treatment of homeownership in tax policies. From the overview of the tax policies in a number of countries in the eurozone provided by Table 2.3, it can be seen that there is not a straightforward relation between the fiscal treatment of homeownership and the homeownership rate. For example, Slovenia which has limited tax subsidies has a high home ownership rate, whereas the Netherlands with a generous tax policy has a homeownership rate below the EU average.

Table 2.3 Interest rate tax advantages, 2008

	Tax on imputed rent	Tax deductibility on interest payments	Capital gains tax on selling own house	Inheritance tax on own house	Transaction tax (stamp duty)
Corporatist					
Belgium	Yes	Yes	No	Yes	Yes
France	No	Yes	No	Yes	Yes
Germany	No	No	No	Yes	Yes
Netherlands	Yes	Yes	No	Yes	Yes
Social democratic					
Finland	No	Yes	No	Yes	Yes
Mediterranean					
Italy	No	Yes	No	Yes	Yes
Portugal	No	Yes	Yes		Yes
Spain	No	Yes	Yes	Yes	Yes
Liberal					
Ireland	No	Yes	No	Yes	Yes
Eastern					
Slovenia	No	No	No	Yes	Yes

Source: ECB (2009)

There has been a general trend of reducing the costs of such programmes, or regrouping the resources and cutting the funding of generous mortgage subsidy schemes. The favourable tax treatment of housing loans was eliminated in Germany in 1986 as a reaction to the introduction of alternative housing subsidy programs. In Britain mortgage interest payment was fully deductible until 1974, and then a ceiling on the size of the mortgage eligible for interest reduction was introduced. Because the ceiling was not indexed, its significance decreased with inflation, while in 1993, the tax rate was capped below the top income tax rate, and in 1999 the tax deductibility was terminated. In France, the preferential tax treatment was abolished between 1991 and 2000 (Van der Hoek and Radloff 2007).

2.4.2.2   Declining Support for Social Housing

Alongside such decreases in support for homeownership has also been a general tendency for shifts in the type and scale of support for renting. At the time of huge shortages (in the 1950s and 1960s) investment in new social housing had priority in the countries of North West Europe (Boelhouwer and Van der Heijden 1992). National governments implemented policies through special institutional structures developed from the beginning of the century: in the UK local governments, in the Netherlands housing associations, in Sweden municipal companies, in France special public–private organizations, HLM companies, for example. In the 1970s and 1980s, deregulation, privatisation and an increase of the private sector took place, as policies shifted away from social sectors towards support of the homeowner.

Oxley and Smith (1996) argue that the attempt to reduce housing shortages ceased to be the primary aim of housing policy in the 1990s. At the same time, there was a high need to cut public expenditures. They see political decisions made to expose the housing sector more to market mechanisms, which reduced supply side subsidies that previously (mostly) targeted social housing development. They observe large cuts in public expenditure, improvements in the targeting of subsidies, and increasing homeownership via indirect tools such as tax relief for mortgages. Dübel (2008) also points out that government-controlled agencies have withdrawn from direct subsidising and lending in favour of provision for private lending practice. An additional factor in the shift towards demand-side subsidies is that these schemes have been used to increase the transparency and effectiveness of subsidies (Katsura and Romanik 2002).

2.4.2.3 Increase of Homeownership

Policy changes intended to reduce government support for social housing, even if some of the tax benefits of homeownership had also been reduced. often went, hand in hand, with the aim of encouraging homeownership, This has been most striking in eastern European countries, where, mass scale privatisation changed the picture, and served partially as a shock absorber during the transition (Struyk 1996). Elsewhere, the expansion of homeownership was supported by a variety of policy means. In the UK, the privatisation programme started in the 1980s with the Right to Buy, which gave local authority tenants the right to purchase their homes with the benefit of large discounts on the market value. The promotion of homeownership for low-income households has also been served by schemes like Shared Ownership and, more recently, Home Buy Direct (a mortgage product) for first-time buyers. Belgium has historically promoted homeownership, and low-income households are covered by an insurance scheme. In the Netherlands, there is a mortgage guarantee for low-income buyers.

Although, as discussed in relation to Table 2.1, in most countries homeownership increased there are two main exceptions. In Finland, the homeownership rate decreased from 67% to 58% between 1990 and 2000. The decrease was largely a consequence of a severe economic recession, and a housing market crash. In Germany the homeownership rate has remained stable at a low level of about 40%; here, Bausparkassen schemes are heavily subsidised, encouraging households to save before they buy and this strong emphasis on saving results in a tendency to buy at a later age.

Despite such apparent connections between policy inputs and market outcomes, the overall relation is not a straightforward one as demonstrated by Atterhög (2005). Using OECD data he concluded that homeownership policies have had most effect in non-Anglophone countries with relatively low homeownership rates, and that whereas Anglophone countries with a homeownership tradition might have reached the effective limit of growth of homeownership, there may still be room for growth in the non-Anglophone counties.

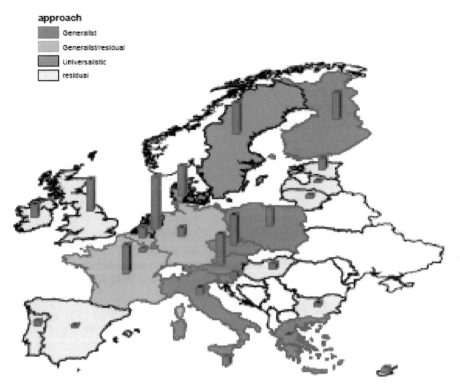

Fig. 2.3 Size of social rental sector and type of approach (Source: Cecodhas 2008)

2.4.2.4   Changes in Rental Housing Sectors

Whereas there are some common tendencies in housing policies, in the main each country has its own policies with few common elements. In European countries, there is a wide range of different social housing 'regimes' (Gibb 2002; Oxley 2000; Priemus and Boelhouwer 1999). There are, for example, different social rental models. In some countries, the social rental sector is residual and meant to serve as a safety net for the most vulnerable households; this is the residual model (Fig. 2.3). In other countries, classified as having a generalist or generalist/residual model the social rental sector is small but intended for a broader target group. Finally, in countries classified as having a universalistic model, the social rental sector is intended for a far larger cross section of the populations. Even more so than with homeownership rates, however, there is not a strong match of social housing models with welfare regimes.

In recent years, the supply of the social housing stock has shrunk in most European countries (Scanlon and Whitehead 2007). A recent development is that countries with broad social housing models have been scrutinised by the European Commission for reasons of competition. As a consequence countries such as Sweden and the Netherlands have been forced to operate without subsidies and cannot target lower income groups to justify state aid. This means that non-profit rental at a below market prices for middle income groups is no longer available, a tenure that was considered by many as an acceptable alternative to homeownership. While this has not been directly intended as a policy to encourage homeownership, it might have that impact (Gruis and Priemus 2008; Elsinga et al. 2008).

2.4.2.5 Household Decision Making

The household interviews carried out as part of the DEMHOW project support the contention in many of the studies reported above that the structure of housing and financial market policies influences the tenure decisions made by individuals. When asked why they had chosen to buy their homes, many responses reflected the financial logic. The overriding picture was that people thought renting was a waste of money – as 'dead money' – and that buying a property was an investment for the future. Some explained that it had been cheaper for them to buy than to rent as well as seeing the potential of building equity and becoming outright owners. Paying a mortgage was seen as a good way of saving money. Mainly, people thought of buying as a good financial investment, only a few specifically mentioned that they bought a house to finance their retirement.

> Buying was a better investment than renting. I am paying but I am making a kind of saving because it is an asset I have. If I rented I would also have to pay and in the end I would have nothing. It is like making a savings account in a bank but in a flat. I am 'depositing' the money of the mortgage. Moreover, the value I pay to the bank is lower than what I would pay for renting a place of the same size. (Portugal, 25–35 years)
>
> I just didn´t want to give away my money to anybody, I wanted to invest in something by myself in order to have something for old age. Pension, I don´t know if I get that someday. (Germany, 25–35 years)

Many respondents tended not to give their reasons for buying but explained that they had an opportunity to buy and did so. In Hungary, for example, buying is what you do when you want a house to live in; there are few rental dwelling available, and buying is considered the only option. Moreover, also in countries where there are rental dwelling available, buying a home 'was just what you did'.

The view that buying a home was the 'natural thing to do', particularly when starting a family, was echoed by a number of households in some other countries, reflecting homeownership ideologies encapsulated in sayings such as 'the Englishman's home is his castle' and the Belgian 'born with a brick in his tummy' (Doling 1997). This suggests that in many European countries there is a cultural imperative, or norm, to buy a home. However, in other countries such as Sweden, Austria and the Netherlands renting and in particular social or public renting is considered an attractive alternative (Elsinga and Hoekstra 2005).

In other countries, the opportunity to buy arose as a result of changes in housing policy following regime change. For example, households in Slovenia explained

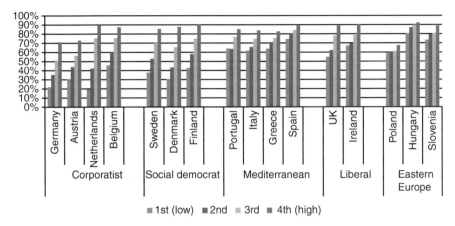

Fig. 2.4 Percentage of owner-occupiers by income quartiles (Source: EU-SILC 2008)

that they had the opportunity to buy their social rented properties as sitting tenants during the transition from socialism after 1992. In Portugal some older households reported that they had the opportunity to buy following the introduction of stricter rent regulations, including the freezing of rents across the country, in the early 1970s. As a result, many landlords left the private rented sector and former tenants were able to buy.

2.4.3   Household Characteristics

Previous empirical research points towards a number of household characteristics, such as family size and educational attainment, being important in explaining homeownership (see for instance Andrews et al. 2011; Clark et al. 1994; Kurz and Blossfeld 2004; Feijten and Mulder 2002).

2.4.3.1   Income

In all countries, homeownership is the majority tenure among higher income groups (Fig. 2.4). In most countries, between 80% and 90% of high income groups are homeowners, the main exceptions being Germany, Austria, and Poland. In contrast, the level of homeownership among lower income groups varies considerably among countries. The difference by income group is relatively small in the Eastern regime countries and slightly less so in the Mediterranean countries. In contrast, the income discrimination is large in Corporatist and Social Democratic countries.

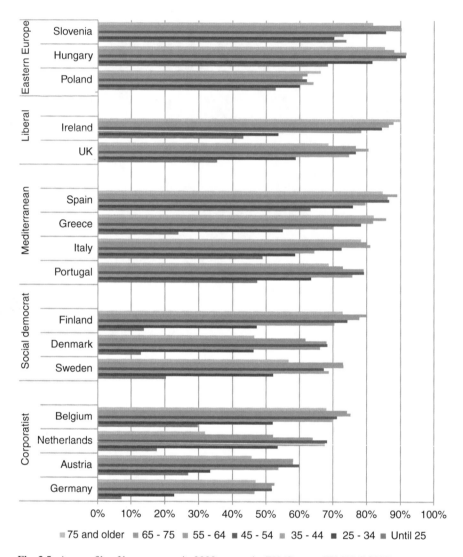

Fig. 2.5 Age profile of homeowners in 2008 across the EU (Source: EU SILC 2008)

2.4.3.2   Age

Figure 2.5 shows homeownership rates by age group. It is not possible to conclude from this whether the percentages reflect age or cohort effects. It is possible, for example, that lower rates for older people reflect the timing of the deregulation of housing finance markets, which had the effect of making access easier for younger generations (Andrews et al. 2011). Nevertheless, there are some distinctive patterns by regime type. The percentage of individuals under 25 who own their own property in

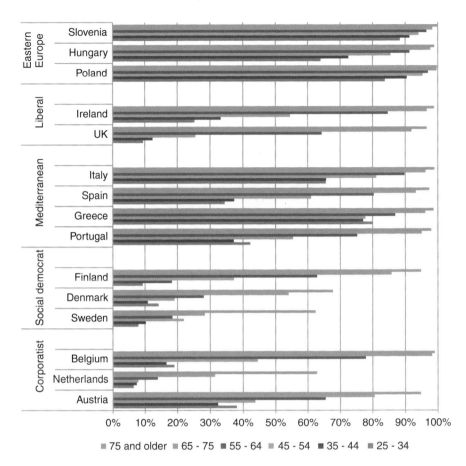

Fig. 2.6 Outright ownership by age (Source: EU SILC 2008)

which they reside is low in the Corporatist and Social Democratic countries, whereas in the Liberal, Southern and Eastern regime countries the percentage approaches or even exceeds 50%.

There are also country differences for older age groups. In the Southern and Liberal countries, their homeownership rates are as high, or almost so, as they are for the middle age groups. In many of the countries in the other regime types, however, the rate is highest for those in the 55–75 age groups, with it being lower for those aged over 75 years.

The rate of outright owners (as a percentage of all homeowners) is presented in Fig. 2.6. The Eastern countries, in particular, but also the Southern countries have high rates of outright owners across all age groups. In contrast, the tendency in countries in other regime groupings is for rates to increase markedly by age group. The differences probably reflect the depth of national mortgage markets. In the Eastern countries even the majority of the younger cohorts are outright owners, whereas in the Mediterranean countries a higher proportion of younger homeowners have outstanding mortgages. Outright homeownership levels appear to be low in the Corporatist countries, reflecting their generally lower rates of homeownership over-all, and in the Social Democratic countries, where even older people appear to have outstanding mortgages.

2.4.4 Combining the Factors

The aim in this section is to bring together the macro and micro factors previously identified, doing so through logistic regression using EU SILC 2008. In this, house-hold characteristics and country context are combined. The results are presented in Table 2.4, which explains the chance of being a homeowner, and Table 2.5 of being an outright homeowner.

Table 2.4 shows that homeownership is related to age, the older households are the higher the chance that they are homeowners. When income is added to the model, the explanatory power increases with the Nagelkerke R2 of 0.05 rising to 0.15. The relation between homeownership and income is as expected: the higher the income the higher the chance that individuals are homeowners. The model also shows that there are significant interactions between these two variables. Households with a low income and a young age have very little chance of being home owner. Low income, older aged people, many of whom will be living on pensions and have earned more in the past at the time at which they become home owner. Therefore, the probability that older, low-income groups are more likely to be homeowners. This provides some confirmation of the relevance of the life cycle model to tenure decisions.

The inclusion of the dummies for the countries also improves the model (Nagelkerke R2 rising to 0.236) and demonstrates that the country context matters with respect to the chance that households are homeowners. The results in Table 2.4 show the deviance from the reference category, UK. Households, in Social Democratic and Corporatist regime countries have a much lower chance of being a homeowner than households in the reference category, the UK, whereas for those in the Southern and Eastern Europe the chances are higher. This supports a conclusion that national, institutional factors such as mortgage markets are important in influencing home-ownership rates. Without it being possible to disentangle the effects of individual factors, these findings are consistent with the earlier discussion related to Fig. 2.1 and Table 2.2.

Table 2.5 provides the results of a similar analysis in which outright homeowner-ship is the dependent variable. This also shows that age and income are significant, as are different welfare regimes. The countries in the Eastern grouping have high rates of outright owners, which is consistent with the low development of their mortgage markets (see Fig. 2.2). In contrast, countries with low rates of outright ownership, especially the Social Democratic regime countries, have highly developed mortgage markets.

Table 2.4 Logistic regression for homeownership (vs. rental)

	Model 1		Model 2		Model 3	
	B	Sig	B	Sig	B	Sig
Age						
(REF)75 and older						
Until 25	−1.500	0.000	−1.050	0.000	−1.273	0.000
25–34	−0.517	0.000	−0.642	0.000	−0.833	0.000
35–44	−0.020	0.316	−0.227	0.039	−0.262	0.020
45–54	0.187	0.000	−0.047	0.670	−0.082	0.466
55–64	0.289	0.000	0.203	0.073	0.156	0.180
65–75	0.196	0.000	0.237	0.067	0.308	0.020
Income						
(REF) highest income quartile						
First			−1.228	0.000	−1.460	0.000
Second			−0.997	0.000	−1.105	0.000
Third			−0.569	0.000	−0.612	0.000
Interactions						
First quart <25			−1.373	0.000	−0.993	0.000
First quart* 25–34			−0.904	0.000	−0.689	0.000
First quart* 35–44			−0.700	0.000	−0.682	0.000
First quart* 45–54			−0.550	0.000	−0.458	0.000
First quart* 55–64			−0.495	0.000	−0.358	0.003
First quart* 65–74			−0.232	0.082		
First quart* 75+						
Second quart* <25			−0.523	0.000	−0.417	0.006
Second quart* 25–34			−0.392	0.001	−0.329	0.006
Second quart* 35–44			−0.386	0.001	−0.469	0.000
Second quart* 45–54			−0330	0.005	−0.356	0.003
REF=UK						
IE					0.379	0.000
AT					−0.958	0.000
BE					−0.219	0.000
DE					−1.293	0.000
NL					−0.680	0.000
DK					−0.652	0.000
FI					−0.206	0.000
SE					−0.422	0.000
ES					0.749	0.000
GR					0.124	0.002
PT					0.070	0.105
IT					−0.017	0.568
PL					−0.443	0.000
SI					0.578	0.000
HU					1.161	0.000
Constant	0.796	0.000	1.843	0.000	2.157	0.000
−2 log likelihood	183,787		171,870		161,515	
Nagelkerke R2 (pseudo)	0.050		0.153		0.236	

*Only significant interactions presented

Table 2.5 Logistic regression for outright ownership (vs. mortgaged owners)

	Model 1		Model 2		Model 3	
	B	Sig	B	Sig	B	Sig
Age						
(REF)75 and older						
25–34	−3.057	0.000	−2.455	0.000	−3.403	0.000
35–44	−3.047	0.000	−2.749	0.000	−3.484	0.000
45–54	−2.476	0.000	−2.161	0.000	−2.667	0.000
55–64	−1.954	0.000	−1.712	0.000	−1.875	0.000
65–75	−0.961	0.000	−0.917	0.000	−1.046	0.000
Income						
(REF) highest income quartile						
First			1.131	0.000	1.167	0.000
Second			0.337	0.034	0.437	0.015
Third			0.116	0.499	0.204	0.298
Interactions (only significant interactions presented)	*No below 25 years category*					
First quart* 25–34			−0.479	0.006	−0.554	0.006
Second quart* 25–34					−0.346	0.068
Second quart* 35–44			0.366	0.025		
Third quart* 25–34			−0.308	0.081	−0.350	0.083
REF=UK						
IE					0.963	0.000
AT					0.582	0.000
BE					0.571	0.000
NL					−1.742	0.000
DK					−0.977	0.000
FI					0.201	0.000
SE					−1.528	0.000
ES					1.041	0.000
GR					2.388	0.000
PT					1.052	0.000
IT					2.172	0.000
HU					2.332	0.000
PL					3.527	0.000
SI					3.636	0.000
Constant	2.887		2.277		1.723	
−2 log likelihood	105,944		103,494		75,236	
Nagelkerke R2 (pseudo)	0.203		0.233		0.532	

*Only significant interactions presented

2.5   Conclusions

Across Europe, the continuing growth of homeownership indicates a convergence in which, at least in statistical sense, it has become the most popular form of housing tenure. Yet, large differences in national homeownership rates persist. The existing literature provides indications of what might underlie these trends.

The political science–housing studies literature has stressed the importance of a trade-off between homeownership and state welfare provision especially in relation to pensions. Essentially, countries with high homeownership rates tend to have smaller state pension commitments. Whereas this is consistent with the framework developed in Chap. 1 in which housing and welfare systems, along with the family, can be seen as substitutable forms of horizontal distribution over the life cycle, homeownership rates only loosely correlate with a welfare regime typology based on Esping Andersen.

The economics and housing studies literature suggests the importance of other factors or drivers. At a macro level, the growth of housing finance market, itself generally an outcome of deregulation has facilitated higher levels of access to the tenure. National housing policies, too, have made a difference and explain why homeownership rates differ so much over Europe. However, it is not so much homeownership policies that make a difference, but more the policies towards renting. Policies supporting social renting in a number of countries, Sweden, Netherlands, Austria, France, explain why the homeownership rate is rather limited.

The importance of formal, national institutional frameworks within which individual households operated was further supported by the evidence from our interviews. The interviews also suggested the importance of informal institutional frameworks: in some countries with particularly high homeownership rates, for example, homeownership was so embedded into the way of life that it had become the natural and normal option. Examination of other evidence at a micro or individual household level indicates other influences, especially income and age.

Chapter 3
Housing Wealth in the Household Portfolio

3.1 Introduction

The focus in this chapter is on homeownership as a financial asset rather than a physical structure or a legal status. Its broad aim is to establish, in relation to other sources of personal wealth, how significant is the wealth embedded in the homes of European households. Quite simply, given that the majority of Europeans are now homeowners, how is this reflected in their wealth portfolios?

Underlying the presentation of existing evidence and the development of new investigations is the standard life cycle model, which assumes that households save money out of their income over their working lives (typically between their 20s and 60s) in order to smooth consumption over their life course, especially when there is a drop in earnings such as during their retirement. Savings may take a variety of forms, some liquid, for example, cash deposits, and others more illiquid, which may include the household's dwelling. According to the LCM, at any point in the life cycle households have several interlinked choices to make: how much to save this year; how much wealth to accumulate for retirement; and what is the balance between different forms of savings? In practice, household approaches to saving will be influenced by many factors. At an individual level, investment will be influenced by preferences on the balance between spending, savings and debt. At a structural level important factors may include the availability of different financial products and the expected risks and rates of return of other forms of investments such as shares. It may also be expected that the wider welfare system, covering in particular the roles of national governments and the family, may place more or less responsibility on individuals and households to meet their own needs. In those countries in which individual responsibility in this respect is higher, there may be pressures to save more in total as well as more in housing.

It may be expected, as argued in both Chaps. 1 and 2, that also important will be access to financial products that enable households to consume and to invest, in advance of cash savings. In practice, in most EU countries, a significant proportion

J. Doling and M. Elsinga, *Demographic Change and Housing Wealth: Homeowners, Pensions and Asset-based Welfare in Europe*, DOI 10.1007/978-94-007-4384-7_3, © Springer Science+Business Media Dordrecht 2013

of households purchase a home during their working years using a combination of savings and loans. Such loans are generally repaid in full or in part by the time of retirement so that their availability influences the profile of consumption, saving and debt over the life course.

Building on these propositions, this chapter examines the role of homeownership in wealth accumulation across the EU. It has two main parts. The first investigates how housing fits into households' wider life cycle planning – what part does housing play in the composition of wealth, what influences this and to what extent do households see homeownership as saving? The broad picture is that households in all countries do indeed recognise housing as an important investment, and for the average older European, housing equity forms the largest single item in their wealth portfolio. At the same time, their investment decision appears complex, not least because housing is both an investment and a consumption. There also appear to be important interactions operating between personal wealth accumulation and welfare spending.

The second part of the chapter presents evidence and analysis about housing debt in the form of mortgages and about the ways in which this may influence the role of homeownership as a financial asset. Important here in understanding national differences appears to be the availability and price of mortgage products combined with quite widespread disinclination to prolong the repayment period longer than necessary. Debt is thus seen widely as part of the household life cycle strategy whereas being in debt is not necessarily considered to be a good thing.

3.2 Household Wealth

3.2.1 How Much Wealth Do Households Have?

Household wealth is often referred to as net worth, this being defined as the difference between the sum of all the assets owned by the household and the sum of all its liabilities. In this context, assets include financial assets such as bank deposits, shares, securities and loans as well as non-financial assets such as housing. They also include money invested in a private pension scheme but not pension entitlements built up through the state. Liabilities are mainly loans, but may also include other accounts payable.

A fundamental problem facing an understanding of household wealth is the lack of harmonised data over even small groups of countries. Few individual countries undertake annual household surveys that measure the entire spectrum of household assets and liabilities. Where they do undertake surveys, many countries have their own unique way of defining and classifying assets and liabilities. Even where there are apparently harmonised data sets provided by international organisations, such as Eurostat, these often disguise inter-country definitional differences.

Boone and Girouard (2002), for example, report data for six countries only – USA, Canada, UK, France, Italy and Japan – while Altissimo et al. (2005) report for seven – omitting Canada and Japan, but adding Germany, Netherlands and Spain.

Table 3.1 Households' stock of financial assets as a percentage of GDP, 2007

	Total financial assets	Deposit/bank account	Shares/equities
Corporatist			
Belgium	271.0	76.7	108.7
France	185.8	55.6	50.4
Germany	188.4	66.9	47.5
Netherlands	280.3	61.1	42.3
Social democratic			
Denmark	237.8	50.0	70.2
Finland	122.5	38.0	54.0
Sweden	174.5	32.5	68.9
Mediterranean			
Italy	240.8		
Portugal	223.5	80.7	86.5
Spain	182.3	69.5	77.1
Liberal			
Ireland	161.7	37.9	41.9
UK	295.9	78.7	46.5
Eastern			
Czech Republic (2006)	77.2		
Hungary	97.6	34.1	36.2
Slovenia	111.7	50.9	41.2

Source: Eurostat

Even on these samples, there are comparability problems. Altissimo et al. (2005: 13) note of their seven countries, that data for Spain and the Netherlands are taken from a different source to the source for the other five countries and that consequently the figures reported 'are not fully comparable and should be only taken as indicative'. Moreover, whereas, these same authors report total household wealth as consisting of financial wealth and housing wealth, Boone and Girouard (2002) add a third element, other wealth.

Notwithstanding these problems, it is possible to build up some broad pictures of the amount and composition of household wealth across the member states. The first step in building this picture is through an overview of households' financial assets as a percentage of the gross domestic product. Table 3.1 shows that in 2007 the amount of total financial assets clearly varies from country to country, with the percentage being four times higher in the UK than in the Czech Republic, for example. In general, the percentage is highest in those countries in which households, as a consequence of the structure of pension systems, hold large private pension savings; this includes the Netherlands, the UK and Denmark. With the exception of the significantly lower total financial assets as a percentage of GDP in the newer member states – the Eastern grouping – however, there is no systematic variation across regime groupings. In all the countries included in the table, the average

Table 3.2 Median net worth
and gross financial assets
2004

	Net worth (000 €)	Gross financial assets as % net worth
Corporatist		
Germany	98.6	17.3
Netherlands	157.9	13.6
France	170.5	8.0
Austria	105.3	5.7
Social democratic		
Sweden	101.9	24.7
Denmark	128.5	20.8
Mediterranean		
Italy	150.4	1.5
Spain	136.1	1.4
Greece	95.3	2.1

Source: SHARE

holdings of bank deposits and of shares, both as a percentage of GDP, constitute large – in most cases the largest – elements of total financial assets.

For households with a head aged over 55 years, the SHARE data provide additional information, although for nine EU member states only. In addition to financial wealth, Table 3.2 presents the median net worth. Whereas there is some variation in the value of net worth accumulated between the EU member states for which data are available, this is small and the median net worth of older households in all cases is roughly at or above 100,000 euros. There is considerably more variation in gross financial assets as a share of net worth. Households in northern and middle Europe countries, with the exception of Austria, have considerably more financial wealth than those in southern countries. Whereas households in the three Mediterranean countries included here – Italy, Greece and Spain – have on average low amounts of financial assets, for older households in all countries non-financial assets, which will include housing, constitute at least three-quarters of their net worth. In the absence of data for additional countries, it is not possible to say whether this reflects a regime effect or, since they have lower GDP per capita than northern European countries, an economic development effect.

3.2.2   How Much Wealth Is Held in Housing?

At least for older households, then, non-financial assets appear to constitute a major share – three quarters or more – of net worth, and much of this may well be accounted for by housing assets. However, data limitations mean that estimates of the complete profile of housing assets – that would provide unambiguous support for such a conclusion – across the member states even at one point in time, let alone over time,

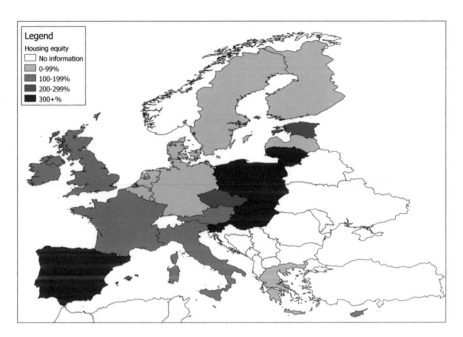

Fig. 3.1 Housing equity as % of GDP (EU25) (Source: Authors' calculations)

are very approximate. Whereas some European countries have reliable measures, annually updated, of house price levels and changes, most do not. For some countries, the Bank of International Settlements publishes an index of house prices by year, but this does not indicate absolute values. For most countries, therefore, there is not a measure of even the average house value. Knowledge of the value of housing wealth owned by individual households is also generally approximate because all homes are unique, if only in their spatial coordinates, with their price being determined individually by the market. Household surveys asking respondents the value of the family home do not necessarily produce accurate estimates.

However, by applying some approximations to available published data from a range of sources, the total amount of net equity held in homeownership, that is house price less any outstanding housing loans, at about 2003 appears considerable, totalling about €17 billion in all member states which is equivalent to about 140% of combined GDP (Doling and Ford 2007). The net equity as a percentage of GDP varies substantially across member states, but there is some match with our five groups: the Mediterranean countries (with the exception of Greece), the Liberal countries, and the newer member states generally have the highest levels of net equity as a percentage of GDP. The Corporatist and Social Democratic countries (with the major exception of Austria and France) tend to have least (Fig. 3.1).

Estimates reported by Altissimo et al. (2005) and Boone and Girouard (2002), again for just a few countries, indicate that housing assets typically constitute between 35% and 60% of gross household assets (Table 3.3).

Table 3.3 Housing as part
of household assets (2000)

Country	Total assets as % GDP	Housing assets as % total assets
Corporatist		
France	440	47
Germany	371	51
Netherlands	479	38
Mediterranean		
Italy	447	49
Spain	521	64
Liberal		
UK	490	39
USA	471	27

Source: Altissimo et al. (2005)

Table 3.4 Percentage of net
worth accounted for by the
home 2004

	Home as % net worth
Corporatist	
France	72
Austria	60
Germany	54
Netherlands	59
Social democratic	
Denmark	68
Sweden	66
Mediterranean	
Italy	82
Spain	86
Greece	85

Source: Lefebure et al. (2006)

These studies also show that the ownership of housing and non-housing wealth is not evenly spread across households. The ownership of shares, which in many countries constitutes a major source of total assets, is not widely distributed among households: in general, in the larger EU economies households in the bottom three quarters of national income distributions own only a small proportion of the total share assets; this proportion in most of the countries being 10% or less. In contrast, homeownership – as reported in Chap. 2 – is more evenly spread, so that in general the 'allocation of property wealth across households appears more evenly distributed than that of financial wealth' (Boone and Girouard 2002: 181).

Because older Europeans have typically paid off most if not all their housing loans by the time they retire, for them net housing equity forms a particularly significant part of their total net wealth. Table 3.4, also from the SHARE data, indicates that for households with a head aged over 55 years, housing constitutes at least half of net worth and in the cases of the Southern countries over 80%.

Although the picture developed in this chapter is lacking comprehensiveness, especially in relation to the Eastern member states, it seems safe to conclude that for the average European, homeownership forms a major share of personal wealth: as Altissimo et al. (2005: 14) put it: 'housing wealth continues to play an important role in all countries'. Housing wealth constitutes by far the biggest single share of net worth across households as a whole, but is particularly important for older households. It is also particularly important in Mediterranean countries where average financial wealth appears to have only a small share of total net worth, though given the limited country coverage of the data presented it is not possible to determine whether both in these countries and those of Eastern Europe there are regime or GDP effects operating.

3.2.3 What Influences the Size and Composition of Wealth?

3.2.3.1 Quantitative Studies

Given the limited availability of harmonised data, it is perhaps not surprising that there has not been the development of a large body of quantitative research investigating and accounting for variations in the composition of household wealth – especially the relative share taken by housing assets – across countries. Moreover, much of what there is focuses not on cross country variations, but on the explanation of changes over time in one country. In these, changes in composition over time are generally presented as a direct consequence of trends in house prices and share prices. Thus, one study reports that 'in the period 1995–2000 financial wealth rose as a percentage of GDP in all countries as a result of strong equity price increases' (Altissimo et al. 2005: 13) while another, that 'the proportion of wealth held in shares has increased to a different extent across countries since 1990.... however, over the past two years, the relative weight of shares has reduced dramatically with the collapse of their prices' (Boone and Girouard 2002: 177).

A few studies have focussed more sharply on the search for explanation of the relative contribution of different asset types in household savings. For our purposes, these studies, too, have limitations. First, as Pelizzon and Weber (2008) point out, much of the work on the composition of household portfolios has concentrated on understanding the composition, not of total household wealth, but solely of financial wealth. For example, Guiso et al. (2002) examine household stock ownership in a number of European countries, noting that over the preceding 10 years this had grown as a consequence of a variety of developments:

> Some of them were transitory, such as the high stock returns in the 1990s, but many are permanent: the privatization of public utilities, the demographic trends, and the growth of mutual funds industry that allowed European investors to acquire diversified positions in stocks at much lower costs than through direct acquisitions (Guiso et al. 2002: 7).

Whatever the value of such insights, they concern a specific and not even, necessarily, a major element in total household wealth, and certainly not one that throws any direct light on issues relating to the relative role of homeownership.

A second general limitation is that a number of studies have analysed a more complete definition of assets, but have empirically focussed on only one country (e.g. Le Blanc and Largarenne 2004; Milligan 2005; Pelizzon and Weber 2008). In such studies, the national context (financial markets, tax systems and so on), though it may develop over time, is a given, so that the analysis does not explore how differences in national contexts structure households' financial decisions.

3.2.3.2   Qualitative studies

Different and in some ways more useful insights into influences on decision making leading to the dominant positions of housing assets in the composition of household wealth are provided by our interviews. It was apparent across all eight countries that decisions about housing reflected its dual nature as both an investment and a consumption good, with some responses stressing one and some the other. The decision to buy a dwelling in the early stages of the life course, for example, was not generally associated with retirement planning, but more as a means of acquiring a decent roof over the head, reflecting a view of owner occupation as a 'home':

> W: The only thing we calculated is the monthly debt and that is not higher than the rent we would have to pay for a dwelling we like. That means a relatively big dwelling with 150 m^2 or so. So for that we would have to pay as much rent as we pay for the credit now.
> M: So for us the function of the house is habitation… the impossibility to lose that living space because a landlord determines the tenancy. Thereby we actually didn't think of an investment.
> (German couple, 65–75 years)

> I see my home as something to live in… for as long as possible… No, no, no other options and… those whatever you name, I just want to live here in peace until death and so…
> (Slovenian, 65–75 years)

> Common sense would say yes [the home is a financial resource], but it is not something that is a big factor. It is not something that consumes us all the time, thinking 'oh it is worth so much now'. Yes, it is a kind of stability, but we are old and when you are old you die, so… it is not something that you are going to be frantic about.
> (British, 65–75 years)

Additionally, when asked whether owner occupation would provide them with financial security in old age, interviewees sometimes rejected the question. Indeed they often saw maintaining the house as so expensive that it could actually be a burden. However, as described in Chap. 2, households often thought of buying as a better option than renting. The latter is, by many, considered as a waste. So although people do not explicitly mention the house as part of an investment strategy, they do see buying a house as not wasting money, but as a wise decision.

At the same time, when talking about old age and pensions, some intervie-wees in all countries appeared to see their home as another potential source of income. This was not always a spontaneous response, but was acknowledged after prompting by the interviewer. In Finland, Germany, Slovenia and the UK, younger interviewees were more likely to envisage the home as a source of income than older ones. In the Netherlands, the self-employed – who, because of the pension system needed to have a more extensive private provision strategy than employees – regarded their home as a particularly important potential source of income in retirement. In Hungary and Portugal it was mostly the interviewees without children who regarded the owner-occupied dwelling as a valuable potential source of income as they had no need to leave the dwelling as a bequest.

Interestingly, when asked what interviewees would have done differently with respect to their financial planning, looking back: in Belgium, Finland, Germany, the Netherlands, Slovenia and the UK, some interviewees regretted that they had not invested more or earlier in an owner-occupied dwelling. In Finland and Germany, it was stated that the financial crisis had shown that housing is a good and stable investment, much more so than investments on, for instance, the stock market.

> To me that [the financial crisis] shows somehow that maybe especially real estate is not such a bad investment. Or inflation, many are talking about inflation; that's even rather positive concerning the mortgages.
> (Germany, 25–35 years)

> Well, maybe, if I would have known what I know now, I would not have invested my money in stocks. I would have bought a house ten times more expensive than this one, twenty years ago. Because since then, it overturned 3 times! If I would have done that, I could stop working now. But yeah, you don't know these things beforehand.
> (Netherlands, 45–55 years)

Investment in real estate appeared a popular, potential way to build up financial means and financial security for old age among the youngest interviewees. In all countries, except the United Kingdom, interviewees said they would like to invest in a second property, while others expected to inherit properties. This would provide a rental income to supplement their pension and, if large sums of money were required, then the property could be sold.

> We plan to, in retirement, live on the management of the patrimony we have. We have controlled things so far and it is likely that our pensions are enough to live comfortably off. Besides we are counting on inheriting some houses that will increase our income. Thus, we are not that worried.
> (Portugal, 45–55 years)

A number of households in most countries already owned investment properties although in Slovenia, Germany and the Netherlands these tended to be the more affluent interviewees or those from affluent families.

These results suggest that housing plays an important role in household strategies. When people are asked why they buy, they do not explicitly mention building assets or pension. However, when exploring old age strategies, housing is often mentioned and in particular the fact that housing expenses are low in old age. Buying a house is mostly not an explicit portfolio decision as was suggested in the previous section; it seems primarily a consumption decision and in the second place an investment decision. One could say that housing is a semiconscious pension strategy for many.

Table 3.5 Stock market returns: 1991–2006

	Mean	Median	Maximum	Minimum	Std. Dev.
Corporatist					
France	5.2	9.6	29.5	−26.5	18.5
Belgium	6.1	5.2	31.5	−18.0	16.3
Germany	3.0	7.3	30.1	−29.9	19.7
Netherlands	5.8	11.1	37.9	−34.5	20.1
Australia	5.5	6.9	17.9	−7.0	7.9
Social democratic					
Denmark	6.5	6.7	33.4	−21.0	16.2
Norway	4.5	4.1	36.5	−26.9	21.3
Mediterranean					
Italy	2.9	10.2	45.3	−26.0	21.6
Liberal					
UK	3.4	6.6	17.9	−20.4	11.4
Canada	5.9	6.4	30.1	−23.9	13.3
USA	6.3	6.6	24.6	−20.0	13.0

Source: Authors' calculations

3.2.3.3   Portfolio Analysis

Following on from both quantitative and qualitative research, is it possible to provide an understanding, at least in the statistical sense, of the variation across countries of the position of housing in the composing of household wealth? One way of approaching this is through the use of portfolio analysis, which begins by considering the investment returns available to households in different countries for two asset types that often predominate in average household net worth: shares and housing. Data available from the OECD for a limited set of European and non-European countries – which do not include any of the Eastern group – over the period 1991–2006 indicate significant differences in equity returns (Table 3.5). Italian, German and UK equity markets experienced low average real returns, being below 4%, compared to those in the USA, Denmark and Belgium, where growth was over 6%. The averages hide the returns for any individual year and most countries have seen returns over 30% in some years; only Australia, the UK, and the USA saw the maximum annual returns below 30% over the sample period. On the other hand, in some years most countries saw a negative return in real share prices of over 20%, with only Australia and Belgium not suffering such a decline. The volatility of the percentage change of real equity prices is given by the standard deviation, and high volatility was experienced in Italy, Netherlands and Norway.

Returns from real house prices, given in Table 3.6, show variations in the increase in the real value of housing wealth. Only in Belgium, Denmark and Norway did homeowners receive average returns of more than 5%, while there were negative returns for German homeowners with returns in Italy averaging only just above 1%.

Table 3.6 Real house price returns: 1991–2006

	Mean	Median	Maximum	Minimum	Std. Dev.
Corporatist					
France	3.5	4.4	12.9	−4.3	5.9
Germany	−1.0	−1.4	2.7	−3.6	2.1
Belgium	5.8	5.1	14.1	0.2	3.2
Netherlands	5.3	4.9	15.7	−2.3	4.8
Social democrat					
Denmark	5.4	5.1	16.7	−2.7	5.4
Norway	4.2	6.7	11.3	−10.0	6.3
Mediterranean					
Italy	1.1	3.5	7.2	−8.3	5.6
Liberal					
UK	3.7	4.7	13.6	−9.4	7.3
USA	3.1	3.6	9.4	−2.3	3.3
Australia	3.9	3.8	14.7	−2.3	5.1
Canada	2.3	1.7	9.1	−5.5	4.4

Source: Authors' calculations

Comparison between Tables 3.5 and 3.6 indicates that, for the period 1991 to 2006, in all countries, except the UK, returns from equity markets exceeded those from owning dwellings. Further, the maximum real returns to the homeowner in any one year were below 20%, considerably below those in the stock market. The negative returns were highest in Norway – at 10% – which, with the exception of Australia, were not so extreme as those experienced in the stock market. As can be seen from the standard deviations, the growth of real house price was relatively steady in each country, with volatility for real house price returns being lower than that in real share prices. Overall, then, housing appears to offer a lower rate of return than shares, but with less risk.

Against this background of returns from the two major asset elements in household net worth, techniques from the theory of finance, specifically portfolio analysis, provide one insight. From this, it can be assumed that, in order to construct the optimal composition of their wealth, households will not just analyse the returns from the various assets individually, but also consider the correlation between the returns from different asset groups. As a rule, they will seek to balance their portfolios such that when one asset group falls in price, another will increase. In other words, there is an expectation that the correlations of asset group prices would be negative. In fact, based on the same data presented in Tables 3.5 and 3.6, for half of the countries in our sample, they were positive (Table 3.7). A possible explanation for this is that household behaviour is influenced by the fact, as indicated by the interview evidence, that housing is widely seen as both an investment and a consumption good. Consequently,

Table 3.7 Correlations: real
stock market returns and the
growth of real house prices

Corporatist	
France	−0.030
Germany	−0.118
Belgium	0.372
Netherlands	0.492
Social democratic	
Denmark	0.653
Norway	0.738
Mediterranean	
Italy	−0.333
Liberal	
UK	−0.266
USA	−0.263
Australia	−0.377
Canada	0.046

Source: Authors' calculations

household behaviour will follow not only from the relative returns from their home but also from their desire to continue (or not) consuming their present housing services. Among other things, this suggests that, while in some contexts standard portfolio analysis may provide useful insights, where housing assets are involved, it does not provide a convincing picture of household behaviour.

3.2.3.4   Regression Analysis

An alternative approach to portfolio analysis uses multivariate regression analysis. Based on the existing literature and the statistical evidence of the cross-country variations of net worth, it is possible to identify a number of variables that might influence the proportion of net worth held in the form of housing (Box 3.1).

The first two variables are intended to capture the effects of changes, in the previous year $(t–1)$, in the returns from shares and housing, respectively. The effects may be direct, that is arising as a consequence of changes in asset prices, but they could arise, instead or also, as a result of adaptations in household behaviour responding to the asset price changes. The variable measuring interest rates is included as this might be expected to influence both the availability of loan finance, which in many countries is critical to house purchase, and the value of equities. Income, measured as GDP, per capita is included in order to establish whether the level of GDP is an important determinant of the overall orientation towards homeownership. Finally, average state expenditure on older people is included to capture the possible impact on saving of the generosity of state welfare provision on portfolio decisions.

Box 3.1

$$LHOUSING_t = b_0 + b_1 CLEQUITIES_{t-1} + b_2 CLHOUSEPRICES_{t-1}$$
$$+ b_3 LINT_t + b_4 LGDP_t + b_5 LSPENDING_t$$

Abbreviation	Variable definition
LHOUSING	Log of net housing wealth as % total net wealth
CLEQUITIES	Change in log of real price of equities
CLHOUSEPRICES	Change in log of real house prices
INT	Interest rates
LGDP	Log of GDP per capita
LSPENDING	Log of average state spending on older people

The analysis uses a rather limited set of countries. Whereas the examination above of returns from shares and house prices was restricted to seven EU and four non-EU countries, the data set available for an econometric investigation of the composition of wealth is even more restricted. The OECD series used provides measures of total household wealth for only seven countries, but it provides a measure of housing wealth for only five – Canada, Germany, France, Great Britain and Italy. This clearly limits the extent of generalisability from the findings. The series is also restricted to the relatively short period 1996–2003. While this also has implications for generalisability, the period is at least one in which expanding economies, low interest rates and deregulated of financial markets would have facilitated behavioural change by households.

Following initial tests of the data and, given the short time series dimension of the data set, the seemingly unrelated regression estimation (SURE) procedure was employed. The restrictions that this places on the error term appear plausible, given the integration of capital markets, the cross-holding of assets and the mobility of households. Table 3.8 presents the parameter estimates and the associated t-statistics. The results are broadly consistent with the literature in demonstrating the expected impacts of changes in equity prices. The negative coefficient on the variable measuring changes in equity prices is significantly different from zero and indicates that, over the period 1996–2003 for the five countries sampled, increasing equity values led to decreases in the contribution of housing assets to overall household savings. It is not possible to conclude from this, however, the extent to which the effect can be attributed to a change in asset prices or to a change in behaviour, with households, in response to rising share values, investing more money in shares.

The coefficient measuring the returns from housing is positive, suggesting that when house prices increase, net housing wealth increases as a share of total net wealth. However, the coefficient is not significantly different from zero so that this cannot be a strong conclusion. There are a number of possible explanations for the lack of significance. It is possible that housing is considered not only an asset but

Table 3.8 Parameter estimates

Variable	Coefficient	t-statistic
LCEQUITIES	−0.047600	−6.588062
LCHOUSEPRICES	0.029317	1.078288
INT	0.004170	2.169258
LGDP	0.574258	10.11738
LSPENDING	0.715603	10.08832
No. of observations	40	

also a roof over the head. As the evidence from the qualitative research indicated, buying a house is both an investment and a consumption decision, and may therefore not be strongly determined by house price increases. A second possibility is that some households respond to increases in house prices and thus to their total wealth by extracting some of the increase using an equity release product. Study of the wealth effect indicates that such behaviour is more pronounced in countries with highly deregulated mortgage markets – Canada and Great Britain in the data set used here (Catte et al. 2004; Kluyev and Mills 2007). Another possibility is that for at least some of the five countries over the period 1996–2003, there was a positive correlation between equity and house price increases.

The period covered by our data was one of generally low interest rates and positive economic growth. The results show a tendency for higher interest rates to result in an increase in the relative contribution of housing assets. This may be a consequence of the general rule that increases in interest rates impact negatively on share prices, so that the shifting composition arises not because households hold more housing wealth but that they own less share wealth. The results also show that with increasing GDP per capita, the composition of wealth shifts towards housing, which may reflect the general long-run relationship between increasing economic prosperity in a country and increasing homeownership rate: with more wealth, more people become homeowners and more people therefore hold housing wealth.

The coefficient on spending on the elderly is significant possibly because of the inter-relationship between individual life cycle consumption–investment strategies and welfare state spending. The coefficient suggests that where older households can expect to receive high levels of support from the state, for example, through state pensions, long term and other health care spending, there will be less need for individual households to accumulate non-housing assets. Contrary to much of the literature about trade-off between homeownership and state spending, our finding suggests that the trade-off may be between non-housing assets and state spending. The basis of this may be that irrespective of the contribution of the state, households will still need somewhere to live and will, for consumption reasons, invest in housing anyway. Further, given that generally housing assets will be less liquid than non-housing assets, the latter may be perceived as more usable for meeting consumption needs, households will accumulate fewer non-housing assets. So, for our five countries over the 1996–2003 period, the generosity of state spending on older people does seem to impact on household investment behaviour. However, the main impact may be on reducing the perceived need to accumulate liquid assets. We will return to this issue in the next chapter.

3.3 Housing Debt

Given the lumpiness of housing and its high costs relative to both average incomes and to average amounts of household wealth, the ability of European households to acquire homeownership is often dependent on some financial mechanism that allows them to consume in advance of saving. For some individuals, particularly in countries with a strong familial tradition, this may be achieved through inheritance or the pooling of the resources of the extended family. For some, however, smoothing is facilitated by financial markets through which loans are made to households which are repaid over time from future income. Generally, such housing loans, or mortgages, are secured against the collateral of the home itself. The expansion of financial markets and mortgage debt across European countries over recent decades is thus particularly important in understanding how European households build up their savings over time.

As shown in Chap. 2, the total amount of outstanding mortgage debt to the households in the EU27 member states is now very large, being approximately equivalent to one third of total or gross housing equity, but with large variation across member state. Figure 3.2, along with Fig. 2.2, shows that mortgage markets are relatively small – below 20% of GDP – in the Eastern member states, and comparatively underdeveloped, generally between 20% and 40%, in the Corporatist regime countries.

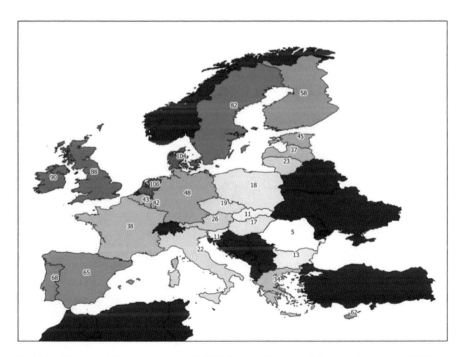

Fig. 3.2 Mortgage debt as percentage of GDP (Source: European Mortgage Federation 2010)

Mortgage markets are much more developed, however, in both the Social Democratic and the Liberal countries.

3.3.1 What Influences the Size of Household Debt?

3.3.1.1 Quantitative Studies

Systematic examination of the factors underlying the variation across member states in mortgage indebtedness has been restricted to a small number of studies. In addition to the (now familiar) problems of the availability or rather lack of availability of harmonised data over a sufficient number of countries and a sufficient number of years are other analytical difficulties. Important here is the fact that, seemingly increasingly, not all mortgage debt is taken in order to save in the form of homeownership or indeed to save at all. The increasing flexibility of financial products means that it cannot be assumed that mortgage debt has been taken out solely for purposes of house purchase. In addition to formal equity release products aimed at older people, in practice some housing equity withdrawal is undertaken at various stages during the ownership of a home, often in order to substitute for consumer credit that is more expensive (OECD 2006). Indeed, this is only one aspect of the possible interactions. Financial market imperfections may mean that households and even businesses are constrained in their borrowing so that increases in collateral which commonly occur through the housing market may increase their borrowing capacity. The argument has been put by Bridges and her colleagues:

> If the borrowing constraint of indebted households is tied to the value of their home, rising housing wealth underpins higher indebtedness by permitting households to increase their secured (collateralised) borrowing. And unsecured debt, such as credit card borrowing, may also be higher if households 'feel' more wealthy as a result of house price rises. Moreover, where credit providers and credit bureaux treat homeownership and/or the value of housing equity as a signal of current and future household wealth, this permits homeowners access in terms of credit that would not be available were they to rent property rather than to own. (Bridges et al. 2004: 3).

Notwithstanding the overlap between housing and other forms of credit, Bridges et al. (2004: 11) indicate for the UK that the 'cycle for consumer credit as a whole matches the economic cycle' while also reporting that different types of credit, for example, credit card borrowing, might behave differently. It does appear, however, that the level of household debt is consistent with the LCM. Debt is highest among young (below 35 years old) and middle age adults, declining sharply among those above 65 years (OECD 2006). Brown and Taylor (2005), on the other hand, found that while this was the case for Great Britain and Germany, the USA pattern was different. Further, in a study of 11 countries in the eurozone, total lending to the privates sector – which was roughly equally split between lending to households and to corporations – was found to be positively related to real GDP and negatively related to interest rates, suggesting that debt increased when economies were expanding rapidly and when the cost of borrowing was low (Calza et al. 2001). Broadly, the argument has been substantiated in a study of 16 industrialised economies.

Evidence for the USA suggests that the demand for mortgage debt is not simply driven by the demand for housing, and is highly responsive to the arrangements enabling mortgage interest payments to be deducted from income tax liability (Follain and Dunsky 1997). The viability of this, as a general rule, that tax systems that alter the after-tax returns of a given investment resulting from national tax systems affects asset selection, is not supported by study of mortgage debt in Italy (Japelli and Pistaferri 2002). Indeed, reform resulting in the withdrawal of mortgage interest relief in that country did not show evidence of any change in the demand for mortgage debt. In contrast, a study of the EU15 member states indicated that the growth over time in mortgage debt was affected by net interest rates, that is, taking account of the impact of mortgage deductibility arrangements (Wolswijk 2008). In addition to tax deductibility arrangements, the latter study demonstrated that increases in national rates of mortgage debt have been positively affected by financial deregulation measures, stock market growth and house price increases (Wolswijk 2008).

Incomplete as they are in providing an understanding of the cross-country variation in the development of housing finance markets, these studies point to a number of general conclusions. First, because the acquisition of homeownership is both a consumption and an investment decision, mortgages provide not only a means of acquiring somewhere to live in advance of having saved the full cost of purchase, but also a means of transferring saving from money income into saving in the form of real estate. Furthermore, as financial products increasingly have been directed not just at the problem of house purchase, based on a notion that people will gain access to homeownership, perhaps in their early working years, and gradually over the course of say 20 years make repayments so that they end up in late working life as outright owners. In other words mortgages do not simply provide a means of smooth financial entry into homeownership. Increasingly, financial products also provide a means whereby housing wealth can be realised. In that way housing finance markets have become vehicles for both saving and dissaving, so that they have relevance to the study of the relationship between housing wealth and income throughout the entire working and post-working parts of the life cycle.

Secondly, on the balance of the evidence from the existing empirical research it seems likely that GDP and interest rates are important determinants of levels of mortgage lending, while the precise role of tax systems including mortgage interest deductibility is less clear.

3.3.1.2   Qualitative Studies

Why Do People Have a Mortgage?

How do households view their decision to take a mortgage? Our interviews show that, as with the decision to become a homeowner, mortgages were often viewed not simply as part of investing for the future but as a necessary concomitant of wanting to consume a certain type of housing. Since very few people in their younger years have sufficient funds to purchase a house outright or have inherited one, a loan was essential. Indeed, in many cases interviewees found the question of why did you finance the dwelling this way rather odd, they had the feeling they had no other choice. A response from a Belgian interviewee was typical:

> If there was another way to surpass the banks, I would certainly do that, but I did not have the money, so I had to lend it. (Belgium, 45–55)

Even where a loan was sought, other financial help was often required. For example, in Hungary very few younger people can buy a house unaided and commonly the family is looked to for help. This help can have different shapes: financial gifts, family loans, a guarantee for the bank and also labour from the family, as a Slovenian example illustrates.

> We got some credits... very little, those that you could get in the past from firms...but mainly with our own money. We were doing construction for about ten years, this means slow... the pace ... so you can build with your own money (Slovenia, 45–55 years).

In general, the help is necessary because in most European countries, the Netherlands being a major exception, financial institutions will lend only a proportion of the value of the home.

Priority Placed on Paying Off Mortgage Compared to Other Priorities

The responses to questioning about the best use of an unanticipated inheritance for someone with an outstanding mortgage revealed attitudes towards indebtedness. None of the 240 interviewees explicitly mentioned using the money for pension building, with only some older people in the Netherlands mentioned a life insurance as a preferred way to spend the money. Most interviewees recommended paying off the mortgage. This was particularly the case with older interviewees for whom having a debt was considered most undesirable. In particular, in Hungary, having a mortgage is considered very risky, something to get rid of as soon as possible, a view that reflects the recent problems that many with mortgages in foreign currency had experienced.

> The priority should be to repay the house as soon as possible and after that have a good life... Because today, being without a loan, I think, that is the best thing... (Slovenia, 25–35 years).

> W: Pay off the house! In either case, first of all.
> I: Why would you advise that?
> W: So that the interest costs are because the payable interests are never as high as the credit rate. So paying off the house makes sense in either case. (Germany, 45–55 years)

> They should pay some of the mortgage off to reduce the burden, it is difficult to save at the moment but they need to look at some sort of saving plan for their children's education. I would not advise them to blow it! Paying your mortgage off is such a boost, it makes such

a difference and they would have more money each month – they should do something
sensible with it…but it is okay me saying that now, would I have been so sensible in my
thirties? (UK, 65–75 years).

However, in some cases paying off the mortgage is weighted against other
options. Low interest rates appear to play a role in Finland and Portugal, countries
where a variable or short-term interest rate is common. The way homeownership is taxed is also relevant here.

My first answer would be repaying the mortgage. But there is a detail here. The best rate the
average Portuguese may get is the rate for housing credit. So, if I decide to use the 30.000
euro to repay the mortgage and then I need them for schooling expenses or anything else,
when I get the loan the rate will be three or four times higher than what it would be if it was
housing credit. Thus, if the person thinks s/he may need the money, it is not worthy to repay
the mortgage. And if they have other credits, then the best is repaying them. (Portugal,
25–35)

F: I would amortize the debt.
M: Yes, I would rather do the same though it is very affordable now with the low interests.
But here one should know of course how much they have money for everyday expenses; are
they having a bit tight or not? Anyway, I wouldn't first go and buy a boat, for instance, with
that sum of money, because then the money would be spent in the boat and their expenses
would grow even many times bigger. But, the most sensible thing to do, as I see it, would
be to pay off the debt in order to release some more playing money.
F: So I do also. But a bank person might say: with 1,65, in no way! Good heavens! (Finland,
45–55)

See if it is profitable to pay back the mortgage, because you can always use the mortgage to
reduce your taxes. (Belgium, 45–44)

I would invest and put aside for emergencies. I would certainly not pay back the mortgage,
else it is no longer tax deductible. (Belgium, 45–55)

Tax advantages for homeowners appear to play a role in Belgium, Finland
and the Netherlands. In the Netherlands only one third of the interviewees consider paying off the mortgage, which undoubtedly is caused by the favourable
tax treatment. In these three countries, households can deduct the mortgage
interest from their taxable income; in the Netherlands, this deduction is without
limits and means that one third to half of the interest payments is paid by the tax
authority.

The outcomes of the interviews demonstrate that in most countries households
have a mortgage because they have no other option. Many households in many
countries emphasise they want to pay off the mortgage as soon as possible. However,
other households explain that if they have money, the last thing they would do would
be to pay off the mortgage. The institutional context obviously plays a role here. In
Belgium, Finland and, in particular, in the Netherlands, it is fiscally attractive to
have a loan. This fiscal incentive appeared to play a decisive role in having and not
paying off a mortgage. This is influenced by the tax policy that turned the inclination
to have no debt into considering mortgage debt as something attractive and part of
a smart strategy. The result of this is that households build less housing equity, since
non-housing assets are more attractive.

3.3.1.3 Explaining the Level of New Mortgage Debt

Statistical analyses of the variations shown in Fig. 3.2 are hindered by the length of
the time series necessary. This arises because the outstanding stock of mortgages
captures the borrowing decisions made by households and financial institutions over
a long period of time; in most countries the average length of a mortgage contract is
some 20 or more years. Lags of a considerable length would have to be employed to
capture the determinants of mortgage demand and supply for the proportion of
mortgages from a particular year in the mortgage stock. Given the short length of
the data set, such an approach would be implausible. In contrast, analysis of the
volume of new gross mortgage lending is not so hampered. Furthermore, the
specification of any variable relating to interest rates is also easier. Financial deregulation has resulted in a multitude of mortgages offered to the public. As mortgages
differ in their details, for example, the duration and ability to have payment holi-
days, the cost to the household to borrow funds for homeownership will differ
according to the specific features of the loan. The European Mortgage Federation
has constructed a variable measuring the interest rate on new mortgages, which
captures the diversity of products in each country.

Box 3.2

$$LMORTGAGE_t = b_0 + b_1 LGDP_t + b_2 MINT_t + b_3 LHOUSEPRICES_t$$
$$+ b_4 LEQUITIES_t + b_4 LOVER60_t$$

Abbreviation	Variable definition
LMORTGAGE	Log of new, real gross lending on mortgages in year t.
LGDP	Log of GDP per capita
MINT	Mortgage interest rates
LHOUSEPRICES	Log of real house prices
LEQUITIES	Log of real price of equities
LOVER60	Log of proportion of population aged over 60

Box 3.2 identifies the variables included in the model to explain new lending The
mortgage interest rate and the house price variables are intended to capture the cost
of house purchase. For its part, the price of equities is intended to capture stock
market growth. The relevance of this is that most mortgage providers require some
form of collateral. There are various forms for households to hold their stock of
wealth, for example, cash in a savings account or shares in the stock market, and the
proportions held depends upon a risk return trade off and the preferences of the
household. However, the only type of asset that is easily observable is the real price

Table 3.9 Real gross mortgage lending

	Coefficient	Z value
LGDP	3.147	4.050
MINT	−0.144	−8.550
LHOUSEPRICES	−0.044	−0.240
LEQUITIES	0.090	1.240
LOVER60	−1.314	−1.380

of equities, and this can also be considered as a measure of the difficulty of obtaining a deposit. Finally, a variable measuring the proportion of the population aged over 60 is included in order to capture the impact of dissaving through equity release products.

The countries in the data set are limited, by the availability of suitable measures, to Belgium, Germany, Denmark, Spain, France, United Kingdom, Ireland, Italy, the Netherlands and Sweden. The individual variables are derived from OECD and European Mortgage Federation sources and cover the years 1988–2007. Initial tests on the data indicate that non-stationary estimation techniques are required. An advantage of the Pooled Mean Group estimator which is used here is that it allows the intercept, short-run parameters and error variances to differ across countries (Pesaran et al. 1999). The long-run coefficients are restricted to be the same for each economy, which may be plausible given the level of financial integration across the European Union by the end of the sample period. In effect, then, the analysis takes the cross-country differences shown in Fig. 3.2 as given starting points, with the variables in the model intended to explain the year-on-year development from those individual country starting points.

The results are presented in Table 3.9. Because the data are non-stationary, as tests of significance the z-values do not have conventional distributions. However, the z-values on the first two variables are sufficiently large, relative to conventional distributions, that their coefficients are clearly significantly different from zero. Higher GDP is associated with more mortgage lending, which is consistent with a general tendency across the countries included for higher national income to lead to more people becoming homeowners and/or moving up the housing ladder.

Interest rates also appear related to new mortgage landing. This is consistent with the liberalisation of mortgage markets in many European countries which has resulted in lower interest rates and a freeing up of the supply of credit to the housing sector. In these circumstances the take up of mortgages has increased with a consequent easing of liquidity constraints on the saving and dissaving behaviour of households. It is therefore significant that we have established a long-run relationship in which national income and mortgage interest rates are significant drivers of mortgage debt.

The impact of the other three variables is less definite. The coefficient on house prices is clearly not significantly different from zero, indicating that there does not appear to be any systematic relationship with total mortgage lending. The positive coefficient on real share prices, while probably not significantly different from zero, suggests that mortgage debt tends to be higher in circumstances where real share prices are increasing faster than is generally the case. This is consistent with the wealth effect: where, as a consequence of large increases in asset prices, households deem themselves to be richer, they may seek to realise some of that wealth in order to boost current consumption (Catte et al. 2004; Kluyev and Mills 2007). For some, the realisation may be achieved through increasing the size of their mortgage loan. However, the fact that the coefficient on the house price variable is not positive and, indeed far from significant, does not give great certainty to this conclusion.

Finally, the coefficient on the variable measuring the size of the over 60 population is negative, and although this would not appear to be significantly different from zero, suggests the possibility that people in this age group borrow less than those in younger age groups. While this would be consistent with the LCM – were mortgages simply vehicles for acquiring assets – the effect may be disguised by any usage, by people in this age group, of mortgage equity release products.

3.4 Conclusions

Notwithstanding the limited availability of even non-harmonised data restricting the identification of the precise situation across European member states, some aspects of the importance of housing in household wealth portfolios can be fairly certainly identified. Most prominent among these is that housing wealth constitutes the largest single form of wealth for the average European household, particularly so for older Europeans. In comparison with the ownership shares, housing equity is both considerably larger and more evenly spread.

While that much is clear, the same data limitations restrict statistical analysis seeking understanding of the size and composition of wealth. Based on data for only five countries for the period 1996–2003, analysis confirms that increases in equity prices are followed by a shift in the composition of household wealth towards equities. The analysis does not identify the extent to which this results from the value of existing share holdings increasing and/or the acquisition of large numbers of shares, that is, a price or a behaviour effect. In contrast, the effect of changes in house prices is less pronounced. The impact of state spending on older people does appear significant, suggesting that higher levels of spending are associated with a lower level of non-housing assets in the household portfolio. The limited data basis should prevent any strong conclusions, but these findings do suggest the possibility that the relationships, discussed in Chap. 2, about a trade-off between welfare expenditure and homeownership, may need to be qualified: the response to a lack of generosity in state welfare spending may not necessarily be the acquisition of more housing wealth but of more wealth in forms that may be more easily realisable which does not support the trade off hypothesis.

The investigation of the role of mortgage debt has been similarly restricted by data availability considerations. But, again, some aspects are clear. Most European households buy a house at some point in their life and for most of them it is the largest investment they will ever make. For most households this investment can only be done with a loan or financial support from the family. There is an interesting picture of the spread of mortgage debt over Europe, with a high level of mortgages in Liberal and Social Democratic countries and a low level in the Mediterranean and Eastern countries. Across the countries included in our analysis, increasing GDP and low interest rates, both factors identified in Chap. 2, appear as important influences on national levels of mortgage debt. Other factors, including the size of the over 60 age group in the population, were less so.

To the extent that the evidence both in existing studies and presented in our statistical analyses, provides a picture – albeit very partial and allowing very limited cross-country comparison – which suggests a broad conformity with the life cycle hypothesis. Households certainly accumulate housing assets throughout their life course to form the major part of their wealth portfolios. For many, using housing loans provide a means of spreading payments. There can be little, if no, doubt that homeownership is, and is seen to be, a major financial asset. At the same time, evidence from the household interviews suggests modifications to this, particularly deriving from the fact that homeownership is both an investment and a consumption which is central to people's lives. Often the accumulation of housing equity was seen not so much as a financial strategy as a welcome consequence of a solution to consuming appropriate housing. Likewise, often mortgages were not seen as part of investing for the future but a necessary concomitant of wanting to become a homeowner. But the perception of mortgages differed in different countries. Indeed in many countries debt is seen as something to get rid of as soon as possible rather than a portfolio decision.

Chapter 4
Housing Asset Strategies for Old Age

4.1 Introduction

Incomplete as it is, the evidence presented in the previous chapter supports a view across the EU of homeownership as being a financial asset. This does not mean that all European households see it that way or that, even where they do, it is necessarily the primary motive driving their tenure and portfolio decisions – the consumption motive is also widely significant. Nevertheless, there is much about the observed housing behaviour, as well as the attitudes, of European households that indicates a consistency with the life cycle model. By the time they reach retirement age, the majority of Europeans have used some of their earned income in order to become homeowners, and with many having paid off any housing loans, housing constitutes the largest single element in their wealth portfolio. Just how large it is appears to be related to the characteristics of other investment opportunities, as well as the strength of welfare state provision in their country.

Both the apparent conformity with the life cycle model and the apparent substitutability between homeownership and pensions suggest that homeowners might well use their housing assets in order to contribute to their income in old age, in effect, to a pension. In economic terms, consumption smoothing during the period of retirement may be achieved, in part, by dissaving some or all of their housing assets.

The aim of this chapter is to investigate the extent to which such behaviour has actually happened as well as the extent to which Europeans see it as part of their life strategies. It starts by describing some of the cross-country differences in pension systems, including households' perceptions and opinions on the adequacy of those systems. The rationale for starting at this point is that it provides an important context for whether households see the need to search for additional income at all.

It continues by establishing, as a background to the investigation, the range of ways in which housing assets can, in principle, be realised. This is complex

J. Doling and M. Elsinga, *Demographic Change and Housing Wealth: Homeowners, Pensions and Asset-based Welfare in Europe*, DOI 10.1007/978-94-007-4384-7_4,
© Springer Science+Business Media Dordrecht 2013

because homeownership may provide an income in kind and an income in cash. Each may be achieved to different extents and by different methods, which together can be summarised in six strategies.

The main part of the chapter assembles evidence from a number of sources – including the existing literature and published statistical information, and our household interviews – in order to identify how significant, in terms of usage, the different strategies have actually been. There are two elements to the broad picture. The first is that of a majority of older Europeans benefitting from low housing costs because of the income in kind enjoyed by outright homeowners. The second is that the dissaving of housing assets – gaining income in cash – is probably less common than the LCM would predict: in reality, Europeans tend to hang on to their housing assets. At the same time, there is evidence that attitudes may be changing so that future generations of older Europeans will be more inclined to use their housing equity to meet their consumption aspirations.

The final section of the chapter provides further evidence of the use of housing assets by older people through an investigation of early retirement. Across Europe, more so in some member states than others, there has developed a culture of early retirement, that is the large-scale withdrawal from the labour market some years in advance of formal state retirement ages. In practice, the equity built up in housing appears to have provided many with the means of achieving their exit from work, both because it provides income in kind and income in cash, the latter occurring, if at all, just a few years in advance of actual retirement.

4.2 Perceptions of the Adequacy of Pensions

4.2.1 Variations in Pension Systems

Consideration of the perceived adequacy of pensions starts from a recognition that each of the member states of the EU has its own unique system of pension provision with its own level of resources going to its retired population. Clearly, then, what is being assessed as being adequate, or not, will be different in each member state.

One typology, often applied to the older member states only, identifies a distinction between pension systems based on Beveridgean and those on Bismarckian principles (Castles 2004). The former describes those systems in which there is a strong orientation towards the pension as a safety net, using means-testing or flat-rate payments to ensure that all citizens are protected from the risk of poverty. Among the older member states, both the UK and Ireland have Beveridgean systems. In contrast, Bismarckian systems relate the level of pension payments directly to the contributions made by workers, thus ensuring that those who had high incomes, and as a result high standards of living, continue to have high standards of living in old age; Germany and France are examples. In this typology, the Scandinavian, social democratic countries – Sweden, Finland and Denmark – form one hybrid, having both Beveridgean and Bismarckian elements, with the Mediterranean countries – Spain, Portugal, Italy and Greece – forming another.

A second typology distinguishes between first-tier redistributive systems that aim to reduce poverty and second-tier insurance systems that aim to ensure not simply poverty prevention but a retirement income that is adequate relative to former earnings (Whitehouse 2007). In this respect, adequacy is often measured by the replacement rate. The pensions which retired Europeans receive are thus dependent on the particular combination of redistributive and insurance objectives and of provider sector that collectively define the overall pension system in their member state. As a broad generalisation, the older member states mainly have targeted first-tier systems provided through their public sector combined with a mandatory second-tier system, also mostly through the public sector and in a majority of cases involving a defined benefit scheme. In contrast, the newer member states, though with less uniformity, tend to focus more on a first tier with a minimum safety net and a second tier, frequently involving the private sector and a defined contribution scheme. Overall, this seems to indicate a stronger social insurance element in the older than in the newer member states.

Figure 4.1 provides, for all OECD countries, measures of the replacement rates as well as distinguishing between public, voluntary occupational and mandatory private pensions. With respect to the total replacement rates, these do not closely match with the five regime groups, though the Mediterranean countries (with the exception of Portugal) appear to have particularly high rates, and the eastern countries (with the exception of Hungary) appear to have particularly low rates. For other groups, there is more variation of the Social Democratic countries; for example, Denmark and Norway have above-average rates, Sweden and Finland, below. Likewise, the composition of pensions – as between public, private voluntary occupational and mandatory private schemes – also does not closely match the five regimes, though again each of the Mediterranean and Eastern groups of countries have internal consistency with both focusing on public pension provision.

The figure also points to variations within countries. In the cases of the UK and Sweden, for example, the systems incorporate both defined benefit (DB) and defined contribution (DC) schemes with those sections of their populations falling into the former arrangement having higher average replacement rates. Moreover, by definition, measures of replacement are tied to the underlying amounts and distributions of income from employment. They do not of themselves indicate whether, in any one country, all groups of society will, on the basis of their pension, have satisfactory standards of living.

A different measure of the standard of living of older people is provided by Castles (2004) who argues that, in practice, the standard of living of retired people in economically advanced countries is a function not only of the money provided by national pension systems but also by other welfare services and state transfers covering, for example, health care and subsidised public transport. He uses the term 'generosity', measured as the sum of all the state expenditure directed at older people as a percentage of GDP per capita. His figures indicate that state generosity tends to be highest in the Corporatist countries and lowest in the Liberal countries, with the Mediterranean and Social Democratic countries lying in between. The newer member states generally have systems that are even less generous than those of the liberal

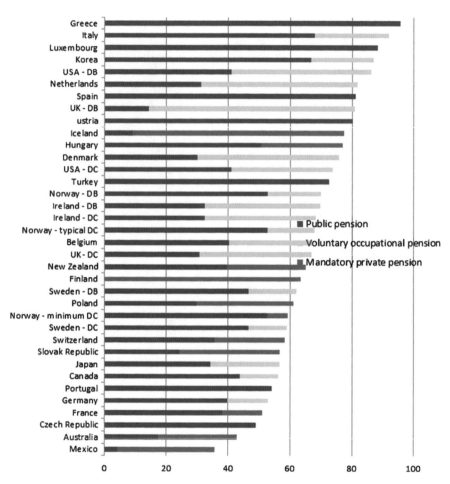

Fig. 4.1 Potential replacement rates at normal retirement age: public, voluntary occupational and mandatory private pensions, 2009 (Source: OECD 2009)

countries. Within-regime-type variations, however, complicate this simple ranking: in the Mediterranean countries, for example, Italy and Spain have significantly more generous systems than Portugal.

4.2.2 Concerns About Pension Adequacy

Whereas state pensions appear to be more or less generous, another issue concerns the extent to which households have ensured that their personal position will be satisfactory. Using a subjective measure of perceived adequacy of income in old age – in the form of the percentage of people who are worried about whether their old age income is sufficient to enable them to live in dignity – Fig. 4.2 suggests that this is

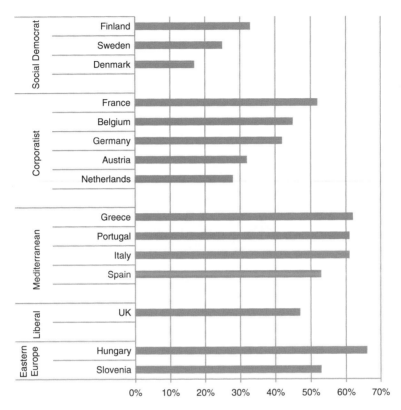

Fig. 4.2 Worries about sufficiency of income in retirement (Source: The Gallup Organization 2009)

often not the case. Insofar as the question asked looks to the future, however, it does not simply describe the reality faced by older people at the present time but can be expected to reflect assessments about likely futures of economies, public spending, political commitments to older people and so on. With the exception of the Mediterranean countries, the country differences broadly fit the Castles ranking: north-west, mainland European countries having lower proportions of respondents expressing concern and Liberal and Eastern countries higher proportions. Nevertheless, for all countries, the proportion of respondents expressing concerns is largely extending; perhaps grossed up, this might apply to as much as a half of all Europeans.

Our interviews provide further evidence of the pattern of concerns about pensions, indicating both a widespread support of, and trust in, the ability of the system in the respondent's country to meet their needs, as well as an assessment of uncertainties, limitations and risks in relying on the state, with a pragmatic view that personal responsibility was essential. In many cases, the dividing line between the first and second views was along generations: younger respondents were generally more inclined to the second view.

Firstly, then, the responses indicated a large measure of confidence in the adequacy of existing pension arrangements. In Belgium, Portugal, Finland, Germany and Slovenia, for example, most of the non-retired thought the public pension would be the most important source of income. In these countries, public pensions are weighted towards earnings-related systems and accounting for a substantial part of the total amount of pension incomes.

In general, the views of most respondents reflected the specific structure of their systems. In Belgium, Finland, Germany and Portugal, interviewees mentioned voluntary supplementary pension schemes as an expected source of income. Interviewees indicated that fiscal advantages made saving in these schemes very attractive. In Finland, additionally interviewees explained that it was an attractive form of saving because the amount was so small that it was hardly noticed. Some interviewees in Finland however appeared 'angry opponents' of these schemes. They did not want to put their savings in the hands of private fund managers but preferred to manage and control their pension savings themselves.

> [S]ometimes when you go into the bank they recommend you invest fifty bucks, they want you to start investing, and for me it feels somewhat pointless since retirement age is so far in the future. I'd rather put the fifty bucks into my own housing, in a way it's a form of investing too. I mean I wouldn't know if I'd ever even get the money. It feels so distant that I don't even bother to think about it.
> (Finland, 25–35 years)

Whereas there was a generally wide confidence in the ability of existing pension systems to meet needs in old age, there were strong indications of differences between older and younger respondents. In Hungary, the UK and the Netherlands, for example, the youngest interviewees typically seemed to have lower expectations of public pensions. They remarked that this would probably be the least important source of income, or they did not count on it at all. In Hungary, this is partly due to the often changing conditions with regard to how the pension is calculated or raised which make the pension system unreliable. A further factor is the general perception that the state pension alone – especially for a single person household – is often not sufficient to maintain an adequate living standard.

> But currently the whole system is so unstable and they say many new things, and many things will change, so I think by the time I will become a pensioner there will be no pension any more.
> (Hungary, 25–35 years)

In the UK, the youngest interviewees hardly mentioned the public pension as a source of income as they expected this to be further reduced over time. In comparison with the other countries, interviewees in the UK were perhaps more aware of the need to make their own provision both because of relatively low replacement rates historically and because of the lack of confidence in state pensions over the longer term.

The evidence from the interviews, then, indicates some, but by no means complete, confidence in the adequacy of the income to be derived from the pension systems established by member state governments. Nevertheless, a frequently expressed view was that governments had a central responsibility to ensure people's needs were met. At the same time, in all eight countries, interviewees believed that individuals had a responsibility to work, to save money and not to live beyond their means and to contribute to various statutory and non-statutory pension schemes. In some countries (Hungary, the UK, the Netherlands, Germany and Portugal), interviewees felt that the state should provide an adequate pension for those who were unable, through no fault of their own, to provide for themselves. For the most part, these interviewees felt that it was up to the individual to save if they wanted to live very comfortably in retirement but that the state should also provide a basic pension for all citizens.

> Every man is the architect of his own future, if he has the adequate income […].
> (Germany, 65–75 years)

Older and middle-aged interviewees tended to see the state as having a more central role, remarking that as they had paid taxes, and made contributions throughout their working lives, the state was responsible for providing them with a pension. Younger people tended to believe that the state should be responsible but explained that it was increasingly unlikely that the state would be able to provide and that individuals would have to take more responsibility in the future.

> I think that responsibility is both from the State and from the persons themselves that should save in some funds. It is just that we got that idea of the public pension for so many years. People are used to that. When I started working, twenty six years ago, putting money aside in private schemes was not an issue. No one thought of that when I started working. There has been a bad management of public money, of social security. And people live longer and longer. All of that contributes for the state of affairs. Thus there is the need to pay longer for the public pension.
> (Portugal, 45–55 years)

In most countries, the public redistributive pension was still considered an important source of income by older interviewees. However, it was evident that younger interviewees overall did not expect the public pension to be as important a source of income in the future.

> I was always aware of it that I preferably blank out the public payments totally and say: 'Okay, I build it independently…' If the state pays you anything else, it will be a bonus and therefore something has changed. I say I don't count on any benefits from the state anymore. Or so minimal, that you shouldn't…yeah, include them.
> (Germany, 25–35 years)

> It will be the money of the PPR (private pension) and the State pension I will be entitled to someday, for the years of contribution. Eventually I mean, if I am entitled to it. There are no certainties, right? If we manage to keep the houses we have, another source of income could come from renting them. But again, there are no certainties.
> (Portugal, 25–35 years)

Younger interviewees expected that saving in personal pension schemes would become increasingly important, as well as building up savings and assets themselves, for instance, in a bank account, by investments in stocks, shares and mutual funds or by investing in housing. Working beyond the formal state retirement age also appeared to be a common feature in retirement plans. In addition, there was a widely expressed view that employers had a responsibility. Younger interviewees in Belgium, for example, believed employers had a role but felt that the state should regulate company pension schemes and provide a guarantee in case the company failed. There were particular concerns about the collapse of company pension schemes in the UK and the limited role that employers could play in providing occupational pensions compared with the past. Although many interviewees felt that, ideally, employers should take some responsibility for their employee's retirement, realistically, the contribution that employers could make was increasingly limited as huge pension deficits made companies uncompetitive.

> This is a difficult one because I know how hard it is – you hear about it in the news.... I do feel that they should provide for their employees but they don't seem to be doing very well these days do they?
> (UK, 60–65 years)

Very few people believed that the family (apart from a partner or spouse) had any responsibility for their income in old age, although people in most countries believed that their wider family would provide financial help or help in kind if they were able to.

4.3 Using Housing Equity in Old Age: Strategies in Principle

Given concerns about the adequacy of existing pension arrangements, how might older homeowners draw on their housing assets? Dissaving of many assets is quite straightforward. In the case of money, it can simply be spent, while many other assets such as shares, bonds, works of art and so on are generally tradable, and once traded, the owner has cash to spend. Housing, at least in the form of homeownership, is less straightforward because of the point, to which our analyses keep returning: it is both a source of consumption and a source of investment. The occupiers of houses receive day by day, week by week a flow of housing services, the amount of those services relating to such characteristics as the size, and quality of the house and its features such as heating systems, as well as the benefits provided by its location and access to facilities and other land uses. The value of this flow of services can be described as an imputed rent which would be equal to the amount of rent the household would have to pay in the rental market in order to obtain an equivalent flow of housing services. Since homeowners do not actually pay rent (to themselves as owners), they can be thought of as enjoying an income in kind from the investment they have made in housing. Because older homeowners have, as reported in Chap. 3, generally completed paying for their homes, this income in kind is received without any offsetting debt repayments.

So, without any act of dissaving taking place, older homeowners receive an income, in kind, from their former investment in housing, which has the effect of

Table 4.1 Strategies for enjoying housing income in kind and in cash

		Income in kind	
		Full	Reduced/zero
Income from equity	Zero	1. Continue to live in home	2. Continue to live in home but let out part to a tenant
	Reduced	3. 'Reverse mortgage product' against a part of the total equity	4. Move down market to a smaller/cheaper house
	Full	5. 'Reverse mortgage product' against all of the total equity	6. Sell home and move into rental tenure

boosting their income. On that basis, literally millions of older Europeans experience homeownership as a pension, while millions of others who are tenants do not. This is strategy 1 in Table 4.1: owners who remain in their homes receive an income in kind proportional to the amount of housing services they previously consumed, but without cashing in any or all of their housing equity.

Without any dissaving of housing equity, homeowners may also use the income in kind in order to obtain an income in cash. Adopting strategy 2, they may let out part of their home, in effect reducing the income in kind enjoyed by them as owners, thereby transforming part of the potential income in kind into an actual rent while at the same time retaining the full equity: housing consumption is reduced, investment retained.

As Table 4.1 indicates, there are additional ways in which pensions can be boosted by homeownership. All involve the realisation of some or all of the equity, and all result in an income in cash. Strategy 4 involves selling the home and buying another, cheaper one, thus realising some of the housing equity. This will also reduce the amount of the income in kind but allow an increase in non-housing consumption. Strategy 6 involves selling the home and renting another, thus realising all the equity. This may or may not lead to a reduction in the total flow of housing services but will necessitate the payment of rent.

Strategies 3 and 5 are significantly different. These strategies may take the form of the use of a financial product, often referred to as a reverse mortgage, or the establishment of an agreement with a third party, which has the general outcome of enabling the owner to continue to enjoy the income in kind from the home, while also releasing some or all of the equity. In some countries, there have long been established legal arrangements whereby money in kind – in the form of care – or money in cash has been given to the older homeowner in return for legal title to the home on the death of the older person. In France, for example, this is the well-known *viager* system. Increasingly, however, there has been interest in reverse mortgages as equity release products, available from financial institutions, which may take the form of an extension to an existing mortgage or a new mortgage altogether, with both providing loans against the collateral of the home to be repaid according to some predefined schedule and usually from monthly income. Alternatively, the product may provide a lump sum or monthly income that is only repaid on death or at the sale of the home; this is often referred to as a reverse mortgage. These possibilities allow the level of housing consumption to be retained while increasing non-housing consumption by drawing on housing equity.

4.4 Using Housing Equity in Old Age: Strategies in Practice

Given the six possible strategies, what actually happens in practice? To what extent are the different strategies pursued by older Europeans? As with other parts of this book, the empirical evidence both from existing studies and collected by the DEMHOW project is incomplete. Nevertheless, it does provide some broad pictures.

4.4.1 Using Non-housing Assets

Any realisation of housing equity of course takes place in a context in which the older European household may not only have pension entitlements but also typically holds a range of asset types in their portfolio. A prior issue then concerns how far Europeans, once they reach retirement, begin to run down their total assets in a way which is consistent with the LCM.

Because the SHARE data covers households with a head aged over 55 years, it can provide a picture of developments over only part of the life cycle. Nevertheless, Table 4.1 indicates that the age profile of net worth – the total value of their assets less the total value of all debts and liabilities – appears to indicate that households build up assets when in work, that is, up the age of 60 or 65, and subsequently run them down to finance consumption during retirement. Such an interpretation is necessarily tentative because the SHARE data does not allow the separation of age and cohort effects. If older cohorts had lower lifetime incomes than younger cohorts, this would result in lower average net worth, quite independently of any life cycle effect. A further complication is that we do not know whether those who die younger have different levels of net worth than those who survive. At most, what can be justifiably claimed about Table 4.2 is that the asset profiles by age groups are not inconsistent with dissaving in old age.

Given this, the between-country variations are worth noting. Although for most nations the peak level of net worth is when the individual is between 55 and 59 years of age, the peak in Austria is at an earlier age and for the Social Democratic countries at a slightly later age. Nevertheless, the age when individuals have amassed their maximum net wealth coincides roughly with their final years in the workforce. As the statutory retirement age for most EU nations is between the ages of 60 and 65, this is the time in most economies when the households might start to deplete their stock of assets that they had built up over their working lives.

A further difference between countries is the age at which households saw the largest decline in net worth. In the Netherlands, France, Switzerland, Germany and Greece, the

Table 4.2 Median net worth by country and age group

Age	Corporatist					Social democratic		Mediterranean		
	FR	DE	AT	CH	NL	SE	DK	IT	ES	GR
<55	181	113	166.1	151.1	195.2	82.3	117.8	160.2	140.2	148.5
55–59	204.5	153.5	116.2	257.2	204.4	107.7	125	204.4	163	167.7
60–64	168	128.8	132.1	250.4	173.5	139.9	145.3	200.1	132.2	135.9
65–69	167.3	118.7	110.5	185.6	83.1	126.2	112.3	154.4	152.3	105.1
70–74	169.7	61.3	76.2	186.1	76.8	86.2	71.4	141.8	116	88.9
75–79	156.3	102.6	73.4	100.5	50.9	70.4	75.4	133.1	123.9	96.8
80–84	125.6	21.6	26	168.7	32.5	54	44.6	63.3	92.8	70.3
85+	95.2	5.2	6.6	81.7	9.6	42.6	38.6	10.8	103.3	48.9

Source: SHARE

biggest drop took place in the years immediately after retirement. However, Sweden, Denmark, Austria and Spain do not follow this pattern as they experienced the largest decrease in net worth when households were 70–74 years of age.

4.4.2   Using Housing Equity

Given the relationship between age and total net worth, it might be expected that housing assets similarly declined for older age groups. It is clear from our house-hold interviews that the possibility and intention of using housing in this way is indeed widespread. They indicate that housing assets are sometimes considered as a personal solution to meeting concerns about the inadequacies of state pension systems. For example, in Hungary, except for the richest households, housing wealth appeared important for financial well-being. Many interviewees can hardly get by in their working lives and consequently cannot save for retirement. In Hungary, retirement does not logically imply that people stop working. On the contrary, many Hungarians will – if they have the opportunity – continue working, and the steady but low state pension will enable them to save, to give more financial support to the children or to consume somewhat more for some time. Then, if households are no longer able to work due to serious health problems, and they are not able to manage financially, they would in the first instance be supported by family members. However, if the family is not able to give sufficient support, the elderly will calculate how much money they could cash if they would downsize. In Hungary, downsizing appears often to be regarded as a realistic option.

In the UK and Finland, some interviewees expressed a strong distrust towards private pension funds; instead, they preferred to invest in bricks and mortar. In the UK, there have been cases where people lost their pension savings and people had become wary of private pension funds. In response, the youngest age group indicated they would prefer to invest in their owner-occupied dwelling and to climb the housing ladder. During retirement, they 'planned' to climb down the housing ladder.

In Finland, it is partly the aggressive marketing for private pension insurances by the banks that seemed to get on young interviewees' nerves. Furthermore, they preferred to stay in control of their pension savings, their financial means, instead of leaving it in the hands of the private pension insurers. They regarded investment in housing as more attractive. Compared to the youngest British interviewees, the youngest Finnish interviewees seemed to speak more about second properties, instead of their owner-occupied dwelling and a housing ladder. They stated that instead of being dependent on a pension fund, they would be in control of the investment themselves.

> Somehow it would feel like owning something concrete, it's something there for me and if I die or something then it would be passed on to my children and they would get to own it. I enjoy owning something concrete. I don't want them [the insurance company]to send me something in the mail every month in order to find out the current rates.
> (Finland, 25–35 years)

In the Netherlands, the self-employed who – in contrast to employees – do not automatically participate in mandatory pension saving, often included their housing in the planning. The retired interviewees who had been self-employed during their working lives had often downsized. The self-employed in the 45–55 age group had plans to sell.

> Man: Yes, the house is our nest-egg. It gives us the luxury to do many things we would like to do. It is the cream on the jelly. We will sell the house, because with the two of us, what would we do with it. It gives us way too much work, six sleeping rooms… With the children it is perfect, and with family staying over. But at some point it will be too big.
> Interviewer: What will you do then?
> Man: No idea, it depends, maybe we will buy a camper. I have no idea really.
> (Netherlands, 45 years)

So, housing plays a role in pension planning, but how should that work? What kind of strategies do households apply? In fact, little is systematically known about the extent to which the six strategies outlined in Table 4.1 have been adopted in practice by older Europeans. Here, we draw on a number of sources organised around three sets of strategies:

- Dissaving by moving (strategies 4 and 6)
- Dissaving but not moving (strategies 3 and 5)
- Not dissaving (strategies 1 and 2)

4.4.3 Dissaving Housing Assets by Moving

Given that housing appears to figure widely in pension planning, the next step is to consider the extent to which housing is actually used as a pension. One possibility is that it is used through strategies 4 and 6, both of which involve the household moving home. The first issue then is whether older European homeowners move at

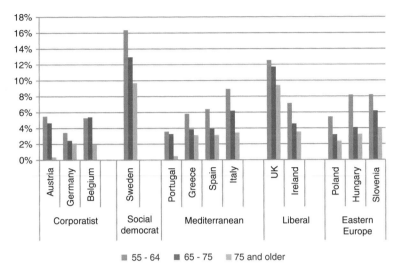

Fig. 4.3 Mobility among all households in older age groups (Source: EU-SILC)

all, a measure of which is provided by age-specific mobility rates. Figure 4.3 shows the share of older households in all tenures that actually moved in the previous 5 years; in all countries, the mobility rate is higher among the 55–65 age group than among the 65 plus group, but the rates vary over countries. In Portugal, people are least mobile, while in Sweden and the UK, households move quite often, almost one-fifth of the 55–65 group moving in the last 5 years. Figure 4.4 provides the mobility figures for homeowners. When compared to Fig. 4.3, it becomes clear that the mobility rates among homeowners are lower than those of all households. Overall, with the exception of Sweden and the UK in the sample of countries for which evidence is presented, mobility rates among people aged over 65 years are low, below 5% in a 5 year period. This suggests that once they reach retirement age, older people become less mobile and in practice most remain for an extended period in their existing home.

There are a number of things that these two figures do not tell us, however. There is no indication of whether households are moving between tenures and, if so, whether this is typically from buying to renting or vice versa. Further, they do not indicate whether households in the 56–64 year age group are typically or predominantly moving in anticipation of retirement and a need to enhance income in cash. Finally, they do not indicate whether moves typically or predominantly to cheaper homes are within the homeownership sector or into the rental sector. In short, they are very partial indicators of the incidence of strategies 3 and 5. Nevertheless, the low mobility rates do suggest that they are not greatly used and that most people die in the homes they bought while they were working.

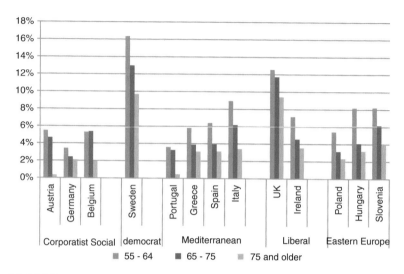

Fig. 4.4 Mobility among homeowners in older age groups (Source: EU-SILC 2008)

In these respects, further evidence is available from empirical studies of mobility behaviour. Bonvalet and Ogg's review of the residential strategies of older people in France identified two periods later in the life course when mobility rates increased: around the age of retirement and above the age of 84 (Bonvalet and Ogg 2008). In the former period, the move was commonly within the home owning tenure, from a larger to a smaller home, apparently adjusting to a smaller income with retirement and/or a decrease in family size with the departure of adult children or early widowhood. In the latter period, the moves were generally triggered by poor health or loss of a partner and often involved a move out of owning to the rental sector including social housing. While the increased mobility at these two periods is consistent with strategies 4 and 6, respectively, rather than dissaving they might also be seen as adjustments to certain life course events, the authors concluding that 'house moves in France by those aged over 75 years are prompted by unsuitable homes coupled with a strong wish to move, either by the individuals concerned or on the part of their family' (Bonvalet and Ogg 2008: 764). These are moves, then, that might be considered to be motivated by desires to adjust consumption rather than to realise investment.

In addition to surveys of mobility, there have been a number of econometric studies based on an LCM framework. Broadly consistent with the Bonvalet and Ogg findings, these indicate a low propensity to realise housing assets in old age. Many have used data for the USA or the UK for the 1980s and 1990s, their results indicating that moving out of homeownership into rental housing, downsizing or otherwise drawing on equity in order to support consumption as families age was not widespread in those decades, an exception often occurring with the death of one partner or a move into a nursing home (Venti and Wise 2001; Rohe et al. 2002; Feinstein and McFadden 1987; Ermisch and Jenkins 1999; Disney et al. 1998).

Chiuri and Jappelli (2006) investigated the pattern of homeownership among older people in 17 OECD countries finding that ownership rates declined considerably after the age of 60 in most countries. However, their investigation concluded that most of the decline was the result of cohort effects, and adjusting for this, rates fell after the age of 70 by only about half a percentage point per year. Other research with a European focus has broadly confirmed these findings while also indicating important differences between countries. A study by Tatsiramos (2006) shows that most European households who are homeowners by their late 50s remain so throughout their remaining working and their retirement years so that there is no evidence of large-scale decumulation of housing assets. Nevertheless, there is evidence of some decumulation as people get older as well as significant regional differences. Particularly where housing costs are high and following the death of a spouse, the mobility rates of older Europeans increase, marking a general, but not large-scale, tendency to move into rental housing or to remain in homeownership but move to a smaller dwelling (Tatsiramos 2006). Older people in the Mediterranean member states (Italy, Greece, Portugal and Spain) are less likely than their counterparts in other western member states to move at all or to move into smaller dwelling or out of homeownership.

The general picture, then, is that older people have limited propensity to move and thereby use their homes as a source of wealth that might contribute to their income needs. Against a background of some country differences in which the Mediterranean countries in particular stand out as low dissavers of housing equity, most dissaving by means of strategy 4 takes place at or around retirement, and most of strategy 6 at 75 plus often following poor health or loss of a partner.

The household interviews largely confirmed these conclusions. However important homeownership is as a financial asset, many people also relate to it in other, non-financial ways. They consider housing not only an asset but in the first place as a home, where people feel safe and able to encounter members of the family (Elsinga et al. 2007). This deters people from moving in order to dissave strategy. In the Mediterranean countries, especially, homeownership has been part of the family ideal in which the owner-occupied home is the physical and emotional focus of the family, where the home is part of the family project, purchased through the combined resources of the extended family to be passed on through generations of the family (Allen et al. 2004). Selling the house and in that way releasing equity is therefore considered not just a portfolio decision but an often emotional one.

The interviews, in a variety of ways, reflected such non-financial dimensions of the homeownership experience and thereby explain the reluctance to move. A common set of concerns related to 'losing' something precious. In all countries, interviewees mentioned the attachment to the 'home'. After discussing the 'irrational excessive saving behaviour' of the current elderly, a Finnish interviewee states the following:

> I understand it much better that one does not want to sell one's dwelling. It is like a lot of memories get into these walls over the years, so that, it can be much more valuable than the bank account... well, it is dear to you, that dwelling. But I can't understand how an account

can be dear to anyone, a bank account. That one does not want to give up the dwelling that
I understand much better.
(Finland, 25–35 years)

In Germany, interviewees emphasised the great attachment to their homes, by
referring to the fact that owner-occupiers put a lot of effort in renovation and sometimes even self-constructed their dwellings. In Portugal and the Netherlands, it was
emphasised that this attachment would be something that would typically increase
with age. The current younger groups could still imagine moving in old age, but
when they would reach their old age, they would probably be similarly attached to
their home as the current elderly.

Interviewees in Germany, Belgium, the UK and Portugal mentioned that consuming housing wealth would mean that all their years of work of asset building
would vanish at high speed. This seemed very unattractive to them, it would cause
an unpleasant feeling. Were they to make use of equity release products, they would
be shifting from independent owner-occupiers into dependent mortgagees. In
Belgium and the Netherlands, interviewees remarked that it would be like renting
again. In contrast, the aim in life would be to become increasingly independent during the life course. Therefore, equity release was felt to be inappropriate.

Some interviewees argued that having savings and having assets represents
power, respect and value. Overall, there seems to be a deep fear to end up with
nothing.

It's the way they [the current elderly] are brought up. It is hard work to save all that money
and they are reluctant to spend it. You become conditioned to your own economy. I think if
you started to spend money – it is easier to start than to stop… you can lose control.
(UK, 65–75 years)

Because it gives such a good feeling to have something. To have some reserves. Very old-
fashioned. Just the feel of it is convenient. And then, if it is not for you, it is for your children. That's important.
(Netherlands, 45–55 years)

4.4.4   Dissaving Housing Assets but Not Moving

The apparent reluctance of older owners to move out of their homes (to move down
market or out of market) may be a function of one or more housing and housing
market attributes. As we have argued, the fact that homeownership is both a consumption and an investment good complicates the dissaving decision. Housing as
consumption may involve considerable psychic or emotional attachment. Established
family and friendship networks and reliance on neighbourhood institutions along
with a store of personal memories attached to the house itself can make changing
one's home of a different dimension to changing the refrigerator or the car. In addition, in some countries, the transaction costs, including legal fees and taxation,
involved in selling a house, and in purchasing another, are particularly high making
the realisation of this form of wealth more expensive than other forms. For example,

transaction costs are relatively low in the UK and Scandinavia and relatively high in Mediterranean Europe (OECD 2004). This may suggest an explanation for the higher mobility rates for the former countries shown in Figs. 4.3 and 4.4. Finally, in some countries, access to decent and reasonably priced rental accommodation may be restricted because of a shortage of supply or of rules of access (Rouwendal 2009).

Insofar as households may view mobility as problematic, strategies 3 and 5 potentially take on added significance. Developments of financial products, referred to here under the general label of reverse mortgages, provide a potential solution because housing equity can be realised with housing consumption continuing as before. Whereas such products have a number of different forms, these can be reduced to two basic types: reverse mortgages proper and interest-only loans.

4.4.4.1   Reverse Mortgages

Reverse mortgages proper provide the owner with a lump sum payment or a regular, monthly income, the amount or amounts paid out being rolled up to be repaid from the proceeds of the eventual sale of the home, which may be at the death of the owner. The pattern of supply of reverse mortgage products varies across member states. From the Reifner report (Reifner et al. 2009; Clerc-Renaud et al. 2010) as well as research by the European Central Bank (ECB 2003), it is possible to identify three broad country groupings. A first group which includes many of the newer member states does not have reverse mortgage products, in many cases because they do not have a legal framework that allows them. In a second group of countries, which includes Germany and France, a legal framework exists that allows the marketing of reverse mortgage products but in practice few suppliers engage with the opportunity and few customers come forward to take them up. The third group has a legal framework, a range of providers and a not insubstantial body of customers. Within this third group, there is considerable variation with a few – Ireland, Spain and the UK – dominating, and indeed with the latter accounting for about three-quarters of the entire European market (Overton and Doling 2010).

Against this pattern of supply, knowledge about usage is limited. With the major exception of the UK, where the trade body collates statistics, there is no central collection of data about both the scale of lending and the characteristics of the consumers or how the money is used. One estimate is that these financial products make up 0.1% of the European mortgage market and hence have not yet become very popular (Reifner et al. 2009). Table 4.3 provides estimates of the total volume of equity release business in a number of the member states. The fact that only some of the member states are included is a reflection of the facts that, firstly, figures are not centrally collated and have had to be independently estimated and, secondly, the absence of an active market in some member states.

Table 4.3 Total equity release sold in 2007

Country	Total volume of equity release		
	Outstanding discounted bill (millions of €)	Average loan value (€)	Number of contracts
France	20.0	100,000	200
Germany	10.0	100,000	100
Hungary	3.2	n/a	n/a
Ireland	n/a	n/a	n/a
Italy	74.3	247,500	300
Spain	1,268.0	352,222	3,600
Sweden	110.0	44,000	2,500
UK	1,825.0	55,303	33,000
Total	*3,310.5*	*83,387*	*39,700*

Source: Reifner et al. (2009)

4.4.4.2 Interest-Only Loans

Interest-only loans differ from reverse mortgage proper in that the homeowner is committed to a regular payment of interest on the loan, and from a normal forward mortgage in that the payments do not contribute to repaying the lump sum borrowed. Sometimes they are offered only with an associated repayment vehicle, sometimes for a fixed period, and sometimes in perpetuity. Not least because the repayment may be made from the eventual sale of the house, interest-only mortgages can fulfil a similar function to reverse mortgages proper (Scanlon et al. 2008). They provide the person who already owns housing equity to extract some of it as a lump sum, which can be converted into an income.

Although, as with reverse mortgages proper there are no centrally collected sources of data covering all the EU, Table 4.4 provides estimates, assembled from a variety of sources, that indicate wide variation. In some countries, such as the Netherlands, they have been popular with older people as a means of consuming housing equity (Scanlon et al. 2008). However, many are taken by younger people, often first-time buyers, for whom they offer a means of accessing homeownership at a lower monthly payment than if they were also paying off capital (Scanlon et al. 2008). This development is particularly interesting in relation to income in old age since, unless the buyer at some point shifts to a repayment mortgage, housing equity is accrued only by house price increase, and an interest-only mortgage in fact more or less turns housing into a consumer good, rented from the lender, with less equity to rely on for old age.

4.4.4.3 Reverse Mortgage Strategies

Notwithstanding significant activity in some countries, in general reverse mortgages of all types, and thus strategies 3 and 5, do not appear in large numbers over Europe as a whole. This matches evidence from the Eurobarometer studies, with Fig. 4.5

Table 4.4 % new loans that were interest only

	1995	2005
Denmark	0	31.5
Finland	0	3
Ireland	7.3	8.4 (12.6 in 2006)
Netherlands (of which	69	87.6 (2006)
no repayment vehicle)	14	44.3
Spain	0	0 (available since 2006 only)
UK (of which no	62	24
repayment vehicle)	10	20

Source: Scanlon et al. (2008)

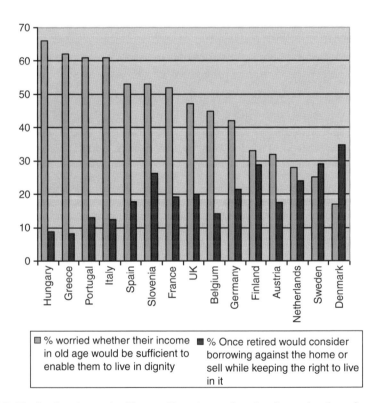

Fig. 4.5 Worries about income in old age and intention to release housing equity, plans of non-retired and actions and plans of non-retired (Source: Eurobarometer 2008/2009)

shows for a set of member states, the percentage of people sampled who responded positively to two questions: were they worried about the adequacy of income in old age and would they consider using a reverse mortgage? Interestingly, there appears to be a significant negative correlation. The earlier discussion about Fig. 4.2

indicated that the first of these two questions appears to be roughly correlated with the generosity of state arrangements for older people. In those countries in which there are the most generous levels of expenditure on old people were also those in which there was least concern about the adequacy of expected income in old age. The responses to the second question are roughly negatively correlated with both. Thus, respondents in Hungary and Greece, in which those interviewed were most concerned about income in old age, were the least likely to consider using a reverse mortgage product.

The proportion of respondents who would consider a reverse mortgage option is highest in those countries which generally have the most developed welfare systems and thereby strongest safety nets and in which, thereby, there is least pressure on older people to have to think in terms of personal solutions to possible future events. The somewhat perverse conclusion to this, therefore, is that the demand for reverse mortgage products, indeed generally any of the strategies identified earlier in the present chapter, may well be highest in those countries where they might be least needed because social protection measures, including pensions, are the most generous.

Even considering the home simply as a financial asset, then, the reverse mortgage decision is a complex one, a view not lost on our respondents, although they also addressed many other issues. In all countries, they maintained the belief that one should not in old age become dependent (again) on a bank: there are risks attached to these types of products, they are perceived as complex, and there appears a distrust towards financial institutions.

> No, I wouldn't consider reverse mortgages. I have bought this house to be able to give my children something. This way, I can give them a starting capital, so that they are able to buy their own house. Even besides that, I wouldn't do that, because your house is no longer your own. It becomes property of the bank. It would feel like renting.
> (Belgium, 25–35 years)

Interviewees also expressed worries about the costs of such products. Slovenian interviewees mentioned high interest rates in comparison to relatively limited gains. In the UK, some of the interviewees thought that reverse mortgages were not good value for money. In Portugal, people feared hidden costs and also in the Netherlands high additional costs were feared. Another issue was the price evaluation by the banks. In the UK and Portugal, some interviewees thought that banks were not trustworthy and underestimated the value of their properties.

Additionally, many interviewees were concerned about the increasing mortgage debt. In Germany, a reverse mortgage would sit uncomfortably with the aim to become an outright owner during the life course, to become free of debt. In the Netherlands and Slovenia, interviewees mentioned an aversion towards incurring new mortgage debt in old age – it would mean losing (part of) the ownership of the 'home'. Furthermore, because longevity is unknown, there is a fear that the debt would grow and could exceed the value of the dwelling. In the Netherlands, Germany and the UK, interviewees argued that this could cause problems: perhaps they would be evicted – which they considered as strongly inappropriate, or their children could be left a debt as a bequest.

However, reverse mortgages were not completely excluded from households' strategies. In all countries, interviewees could imagine using a reverse mortgage as a last resort. Generally, households without children in all countries, except Slovenia, could more often imagine considering this financial, while in the Netherlands, Finland, the UK and Germany, even some interviewees with children could imagine it. In the Netherlands and Germany, the younger age groups seemed more open to the idea of a reverse mortgage than the oldest.

4.4.5   *Not Dissaving*

Much of the evidence, then, suggests that not dissaving housing assets is the most frequent reality for older households in Europe. Taking an LCM perspective, this behaviour might appear irrational. This section provides some understanding, based on our interviews, of the reasons for this behaviour

4.4.5.1   Housing Equity as a Precaution

Notwithstanding the widespread view that housing could be an important part of pension planning, an underlying idea seemed to be that this might take the form not of regular additions to income but as an emergency fund, saving now against future financial shocks. This is additional to LCM behaviour: 'Uncertainty about the life-span, about health and health costs, and the extreme unpleasantness of poverty in old age, combine to make older people extremely cautious about running down their assets. Such behaviour also explains, at least partially, the important role of accidental bequests in the transmission of wealth' (Deaton 1992: 192).

Our interviews revealed the importance of homeownership as constituting a precautionary fund, which people would want to keep as long as possible. People do not know what kind of occurrences they will encounter in life. In retirement, many people have to get by on a lower income than their income from work and therefore feel the need to be prudent and careful. Old age brings health risks, which could potentially pressurise the financial situation. Interviewees mentioned potential changes in social security and changes in their welfare provision, which could have a negative effect on their future financial situation.

> They might be frightened of losing the house. People want to be able to pay for things and not be a burden. They don't know how long they are going to live. And perhaps as you get older, you get more cautious. I think people also tend to think they are poorer than they are. How do you know how much money you will need?
> (UK, 45–55 years)

For many, then, homes are valuable nest eggs. In the event of financial urgency, and without other means to solve the situation, they could cash their housing wealth. Most often interviewees had no precise occurrences in mind that could be the cause of the financial problems; nevertheless, many viewed selling as the most appropriate way to do so.

W: The certainty of having your own dwelling and hopefully no longer having a mortgage.
M: Also the fact that you no longer need to pay rent. And of course, the option to sell it. But
then only in extreme need, when no other options are available. It will certainly not be for
example to afford extra luxury.
(Belgium, 25–35 years)

I think it is more some sort of safety-net kind-of-thing. It [owner-occupation] is not really
something I plan to earn my money with, through purchasing and selling. Because you
normally buy something more expensive after, because you want to live comfortably. But
of course it is convenient to have invested your money in the dwelling. And to know that if
something might happen, you can sell the dwelling and purchase something cheaper and to
have a whole lot of cash available.
(Netherlands, 25–35 years)

In Slovenia, the UK and Belgium, interviewees thought that housing equity
might play a role in their financial strategy if they faced extremely high costs for old
age care. Also interviewees without children in Hungary and Portugal thought that
housing equity might play a role in paying for care. While, commonly, children
may take care of parents, both parents and children may need to balance what
weights heavier: the burden of giving care or the burden of paying for care. Slovenian
interviewees expected to need their pension incomes, children's contributions and
additionally the proceeds from selling their property to be able to afford institutional
care provision.

Why would it [the owner-occupied dwelling] be significant [in old age]? Well, if I happened
to want to go into a home for the elderly, I would probably need to pay more than my
pension would cover, so I might make a deal with somebody for such an additional
payment, otherwise I would need to sell my home. So my home is a kind of security, it is
an income to cover the additional payment for a home for the elderly, If I were to accept
this. I don't know now, one can never know. It all depends upon the circumstances.
(Slovenia, 45–55 years)

Also, in Belgium, a move to a serviced flat or nursing home was regarded as
costly. Some interviewees mentioned that the proceeds of the sales of their property
could partly be used to pay for care.

The house just represents capital and if we want to move to a service flat, then we at least
have budget to our disposal.
(Belgium, 45–55 years)

In Hungary and Portugal, interviewees without children regarded their housing
wealth as a crucial emergency fund. In these two countries, care is still commonly
provided by children, and long-term institutional care is perceived as rather expensive.
In Hungary, monthly care fees are calculated on income and assets, including
housing assets. In addition to the regular monthly payments and an entrance fee,
older people may need to pay extra sums in order to receive so-called 'higher-level'
care. In many institutions, there is the practice of 'parasolvency', which refers to
patients making 'under-the-table' payments to care providers in order to get good
care. In short, having financial means available if one has no children is crucial for
good professional care in old age in Hungary. Interviewees rather choose to use
their housing wealth to pay a caregiver.

It will be important. If something happens, in terms of health, it is possible to sell and use the money to get support of someone, somewhere. Because I think that the prices of institutions for the elderly are absurd. They are absurd, not minimally compatible with the amounts of pensions.
(Portugal, 25–35 years)

4.4.5.2 Housing Equity as a Bequest

The bequest motive provides another addition to the LCM since it implies that individuals may not seek to reduce their wealth holding in old age but rather to hold on to or even to increase it in order to benefit their heirs. Somewhat confusingly, empirical examination of the rate of dissaving of older people – at least in the USA – indicates that starting from the same level of resources, older people without children do not dissave more than their counterparts with children (Hurd 1990; Chiuri and Jappelli 2006).

In our interviews, a persistent theme particularly among those with children was that while a pressing need might lead them to realise some or all of their housing equity, in general this was an inappropriate way of increasing their own living standards. In fact, in all the countries, leaving housing wealth as a bequest seems a common thing to do and is often considered as self-evident.

I would not do that [using mortgage equity withdrawal]. We bought this house after our child was born. I can die tomorrow but my son will have a place to live. I think that someone without children thinks about it. Those who do have children do not.
(Portugal, 45–55 years)

I have bought this house to be able to give my children something. This way, I can give them a starting capital, so that they are able to buy their own house.
(Belgium, 25–35 years)

In all countries, interviewees suggested that older people take into account the financial futures of their children when deciding upon the consumption of housing wealth. In Hungary, housing wealth appeared crucially important for the future of the children. Solidarity among family members is generally expected, as people cannot rely on welfare provision as a safety net. Additionally, the transfer of dwellings is common practice so that consumption of housing wealth would have a great impact on the traditional ways of asset redistribution among family members. Also for German interviewees, leaving a bequest appeared important. They feared social decline in the future that could negatively affect the financial circumstances of their children.

Better you save too much than too little and then suddenly you don't have anything left in old age. You still can pass it on to your children to build up an existence for them. I think that's also a little bit like thinking of the children and saying: I don't need that much money in old age anyway. I don't have that much anymore, I don't travel, don't go to cinema, so what do I need the money for? I rather give it to my children.
(German, 25–35 years)

Notwithstanding the clear significance to many older people of holding on to housing wealth in order to provide a bequest for their children, dissaving and bequests are not necessarily mutually exclusive (Toussaint 2011). Moreover, results presented by Overton and Doling (2010) suggest that also the equity release and family support strategy can be combined. Their results indicate that a significant proportion of households had taken reverse mortgages in order to pass on a bequest to their children in advance of their own demise. In circumstances where people may reasonably expect to survive until their late 80s, their children may themselves already be retired when they do so; a reverse mortgage provides a way not of spending an inheritance but of passing it on when it is most needed.

4.4.6 Changing Attitudes

One set of insights coming from the material we have reported in this chapter concerns intergenerational differences. Much of the existing literature maps the dissaving behaviour of older people using historical data, mainly from the 1980s and 1990s. By definition, these data do not enable the identification of contemporary trends, one of which may, based on research in the UK and Australia, be a growing schism in the attitudes of younger and older generations. In recent years, the practice of SKIING (Spending the Kids' Inheritance) has gained some popular recognition, arguably reflecting a new reality in which the bequest motive has become less dominant. Moreover, it is possible that, as more people reach old age without having had children, questions of the home as a bequest and the significance of intergenerational transfers and solidarity take on a different hue: some might want to leave a bequest to do good things for fellow men, or even the stray dogs of the neighbourhood, but evidence from the UK indicates that increasing numbers of people want to spend it on themselves (Rowlingson and McKay 2007). Australian evidence on this is particularly interesting in that it indicates a clear distinction between the 50–65 and the 65-plus cohorts in that the latter hold on to the notion of the house as a bequest for their children while the former are very much more determined to ensure a continuation of the quality of life achieved during their working years, if necessary by using their housing equity (SEQUAL 2008).

Evidence that this intergenerational schism is a feature of many European countries is presented in Fig. 4.6a, b. Of course, neither shows large proportions of populations stating an intention to use housing as a means of funding their old age, but there is nevertheless a large generational difference with younger generations being two to three times more likely to indicate the intention to use their home in this way.

Aspects of such trends were also apparent from our interviews. In many countries, it was perceived that older people have instilled in them the habit of saving, this being part of their culture, of their socialization. Thus, in Slovenia, interviewees mentioned that older people are more used to modest living than younger people. In Finland, Germany and the Netherlands, concepts such as Protestantism, Lutheranism and Calvinism were brought up. They all referred to beliefs that saving is 'good' and having debts is 'bad'. A general aim during a life course would be to become debt-free once one reaches old age. Not only would using an equity release

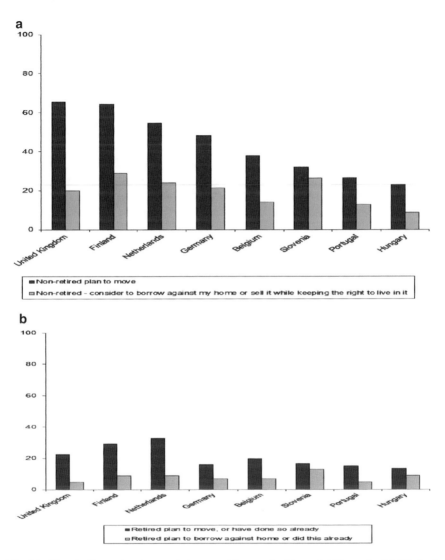

Fig. 4.6 Plans to dissave housing assets to support income in old age: (**a**) Non-retired (**b**) Retired
(Source: Eurobarometer 2008)

product run contrary to this common belief, but many interviewees described their
use as inappropriate luxury, or excessive spending.

However, in all countries, except Hungary, it was often argued that the younger
people thought rather differently to older people. In Slovenia, the youngest age
group was described as cynical about the non-spending behaviour of older people;
for their part, the Finnish interviewees thought the oldest age group should
enjoy and spend more of their savings as 'they won't have pockets in their shroud'.

In most countries, it was suggested that younger people have generally experienced more prosperity and perhaps therefore would more easily spend their savings and assets in old age. Older people instead have often experienced financial hardship. In Belgium, Finland, Germany, the Netherlands and the UK, interviewees referred to the effects of the Second World War on people's expectations.

> People who have experienced the war are scared of never having enough. Social welfare was not back then what it is now. It is stuck in people's brain, also in mine. It is truly stuck.
> (Belgium, 65–75 years)

In the UK, interviewees additionally mentioned the collective memory of historic policy responses to poverty in England. These led to very harsh treatment of people who lacked the means to support themselves. In Portugal, interviewees referred to more recent times of deprivation during the dictatorial regime (1926–1974).

> Before the 25th of April [of 1974 – end of dictatorial regime] there was nearly no pension system. Civil servants had one but not the others. Thus people got used to save, to build assets in order to face any emergency. Because social security was terrible. Terrible or even nonexistent.
> (Portugal, 45–55 years)

4.5 Homeownership and Early Retirement

An indirect way of gaining an understanding of how older Europeans have used, or not, their housing equity in order to support their income needs is through investigation of early retirement. Under existing social protection systems developed in all member states, there is an age, usually between 60 and 65 years, at which citizens are eligible to draw on state pensions enabling them to retire from work. Particularly from the 1980s, however, large numbers of European workers have retired in advance of the formal retirement age which begs questions about the source of the income to support their consumption and the possible ways in which housing fits in with this.

A study of 18 OECD countries, which included most of the western European member states, has provided relevant insights (Doling and Horsewood 2003). At the macro- or national level, the findings show that the coefficient on the size of the homeownership sector was negatively related to labour market participation of older males and significantly different from zero, supporting the proposition that homeownership provides older people with an income in kind, thus enabling them to get by on a lower income than if they were renters. Having equity in the house helps them to withdraw from the labour market before the formal age of retirement.

Further, early retirement was also higher in those countries where house price inflation had been higher suggesting the possibility of a wealth effect on labour market participation: as a result of large increases to their wealth portfolio arising from gains in the housing market, some people appeared to decide that this will support their planned future consumption patterns even without an income from work. These results are based on the macro-data and show a statistical relation but do not allow any conclusions on the micro-level mechanisms behind this. The SHARE database however does provide micro-data and enables us to explore relations between early retirement and different household characteristics.

Our model tests the hypotheses that early retirement decisions are related to the health of the individual, their family circumstances including their marital status and the presence of dependent children, their state of indebtedness and housing factors, including tenure and the length of time they have lived in their present house (see Box 4.1). The results (see Table 4.4) indicate that variables measuring whether the individual is living with a spouse or partner, or is widowed, all have positive and significant coefficients, whereas the coefficients on the separated or divorced variables are not significant. An interpretation of this is that marital breakup often involves a splitting of accumulated assets so that one or both partners may have insufficient investments to maintain their desired consumption levels were they to leave work early.

Box 4.1

$$\begin{aligned} \text{EARLY RET} = b_0 &+ b_1\text{MARRIED} + b_2\text{DIVORCED} + b_3\text{SEPARATED} \\ &+ b_4\text{WIDOWED} + b_5\text{HEALTH} + b_6\text{LONGTERMILL} \\ &+ b_7\text{HOWNER} + b_8\text{HOWNER*Y} + b_9\text{HOWNER*Y2} \\ &+ b_{10}\text{CHILDREN} + b_{11}\text{GRANDCHILD} + b_{12}\text{HHDSIZE} \\ &+ b_{13}\text{DEBT} \end{aligned}$$

Abbreviation	Variable definition
EARLYRET	Not working before formal retirement age = 1
MARRIED	Married = 1, otherwise = 0
PARTNER	In partnership = 1, otherwise = 0
DIVORCED	Divorced = 1, otherwise = 0
SEPARATED	Separated, living alone = 1, otherwise = 0
WIDOWED	Widowed, living alone = 1, otherwise = 0
HEALTH	Good health = 1, otherwise = 0
LONGTERMILL	Long-term ill = 1, otherwise = 0
HOWNER	Homeowner = 1, otherwise = 0
HOWNER*Y	Homeowner * no. years in present home.
HOWNER*Y2	Homeowner * no. years in present home squared
CHILDREN	No. of children
GRANDCHILD	No. of grandchildren
HHDSIZE	No, people living in home
DEBT	Size of debt in €

Table 4.5 The role of household characteristics in early retirement

	Coefficient	Significance
Constant	−0.317	0.310
MARRIED	0.515	0.082
PARTNER	0.568	0.082
SEPARATED	0.074	0.827
DIVORCED	0.009	0.974
WIDOWED	0.486	0.059
HEALTH	0.151	0.016
LONGTERMILL	0.241	0.000
HOWNER	0.071	0.738
HOWNER*Y	−0.022	0.001
HOWNER*Y2	0.001	0.000
CHILDREN	−0.126	0.000
GRANDCHILD	0.131	0.000
HHDSIZE	−0.384	0.000
DEBT	−0.653	0.000

The household size and structure variables may also be correlated with accumulated investment levels and with that their possibilities to retire early. There is a negative relation between the number of children and early retirement and a positive relation with the number of grandchildren. Large households including those with children living at home may well experience ongoing heavy commitments on income that keeps them in work, as well as having depressed their ability to save. This would be consistent with the negative and significant coefficients. On the other hand, the presence of grandchildren will often be an indicator that children have grown up and left the family home reducing the income needs of their parents for consumption and increasing the ability to have accumulated assets.

While Table 4.5 indicates that there is not a significant relationship between early retirement and homeownership, there is with length of stay: the negative and significant coefficient on the number of years that the individual had been a home-owner of their present house and the positive and significant coefficient on the number of years squared indicates a U-shaped relationship. This indicates that individuals are more likely to have retired early in two circumstances: shortly after moving and after having lived for a long time in the current dwelling. The first indicates that realising equity by moving may have played a role in the decision to retire early and is consistent with strategy 4 (Table 4.2). The second indicates that people who have been in their present home for many years, and perhaps having repaid any mortgage and experienced long-term growth in housing equity, providing them with considerable housing income in kind (strategy 1) and the ability to take out a reverse mortgage (strategies 3 and 5). So, whereas over all older people these strategies are generally pursued by a minority, they may figure significantly among those older homeowners that retire early.

4.6 Conclusions

Levels of state spending enjoyed by older people vary across member states in a way that broadly maps onto the five welfare regime types: older people in corporatist countries receiving the most generous levels, with those in Liberal, and especially the eastern countries the least generous, and those in the Mediterranean and Social Democratic countries located somewhere between the corporatist and the liberal levels. Concerns about the adequacy of the different systems appear to broadly reflect not only the specifics of the systems as well as their generosity but in all countries are expressed by substantial proportions of their populations.

Whereas it appears to be widely seen that governments have in practice, and should have in principle, a central, if not predominant, role in ensuring people's income needs in old age, there is, at the same time, a recognition that individuals need to accept responsibility for their own futures. It is against this context that it also appears to be a widespread view that housing in the form of homeownership is a vehicle for pension planning and element in the asset portfolio.

In general, European households appear to reduce their total wealth progressively through their retirement years, in effect enhancing their pensions and in that way acting consistently with the LCM. The dissaving of housing assets, in contrast to most categories of wealth in individual portfolios, however, is complicated because housing is both a consumption and an investment good. Any realisation of housing equity therefore has a consequent impact on the flow of housing services, which in any case, if homelessness is to be avoided, has to be maintained at some level even if all housing equity is realised. Depending on specific national and individual circumstances, most households will face a number of potential strategies although these can be divided into two groups. The first is the group of income in cash strategies, where housing equity is turned into additional income in old age. This can be done in two different ways, by moving or by mortgage equity release. The second group is income in kind strategies in which households enjoy the income in kind in the form of low expenses, in old age.

Moving after retirement is more common in north-west Europe, than in the east and south of Europe. It is not clear whether equity release is the reason for moving; however, there is evidence that moving and early retirement are related, but the numbers are small The numbers pursuing dissaving strategies in the form of reverse mortgages or other financial products appear even smaller. Households are reluctant to buy these products because they distrust financial institutions, they do not know how long they live and they do not want to be independent on financial institutions again.

The most applied option however appears to be that of not dissaving housing assets at all. One question is whether this constitutes a rational strategy. The interviews indicate that households describe two groups of non-dissaving strategies. Housing equity is considered as an element in the family strategy and helps to smooth income and expenses over the life cycle; for example, young people are supported by the family in buying their home with a bequest or otherwise. This demonstrates that although households do not dissave, they do see housing equity as a family vehicle to smooth income and expenses over the life cycle. Another strategy is enjoying the income in kind. At the same time, this housing equity plays the role of a safety net. Home and housing equity appear to have an emotional dimension referring to basic feelings of security in an insecure world. This emotional dimension might be considered irrational from an LCM perspective. However, not releasing housing equity does not mean that housing equity is not relevant for household strategies in old age. On the contrary, housing equity, also without being released, appears to play a key role in household strategies for old age.

Whereas many of these aspects of the tendency to hold onto housing assets vary from country to country, often in patterns which broadly map onto the five welfare regimes. This suggests that the more insecure households are about their income in old age, the less they plan to release their housing equity and want to keep this as a final safety net. There also appears to be a variation, common to all countries, in generational attitudes. The pre-baby boom generations with memories of austerity and hardship were often very cautious, reluctant to spend their assets on consumption and eager to pass on an inheritance to their children. Commonly, baby boomers and later generations' appear much more willing, sometimes eager, to continue, if not increase, the level of consumption they had enjoyed while working, and if this could be achieved by using the equity in their home that was acceptable. Of course this might also be an age rather than a cohort difference, but it indicates the possibility that past attitudes and behaviour may not simply roll into the future.

Chapter 5
Homeownership as a Pension

5.1 Introduction

This chapter is linked with the next in that they are jointly a response to the third of the general questions addressed in this book. Specifically, its aim is to evaluate how the equity embedded in the homes owned by European owner occupiers would, if systematically used for the purposes of contributing to their income in old age, perform as a pension. In other words, it starts from the theoretical premise that in old age European households would fully use all of the equity they had built up in their home during their working years, running those housing assets down to zero, leaving no bequests at the time of death and having nothing in reserve as precautionary saving. Such behaviour clearly contrasts markedly with the actual behaviour demonstrated by European households (as shown in Chap. 4), but it takes the principle of housing-asset-based welfare to a logical conclusion, and through this explores questions of what would be the pension outcomes.

The first step is the estimation of the amount of housing income in kind and in cash that the average household in each member state could potentially enjoy. As in earlier chapters, the analysis is heavily constrained by the availability of appropriate data. This limits the list of member states from which estimates can be derived and the accuracy of those estimates. Our strategy has been to start by using macro-data sets that enable the derivation of somewhat crude estimates for all or many member states. The first two sections of the chapter, which draw heavily on Doling and Ronald (2010), present estimates of housing income in kind and in cash. The second step involves taking these estimates and evaluating them against two criteria: adequacy and sustainability. This constitutes the third and fourth parts of the chapter.

From these two steps, the chapter arrives at a number of conclusions about the potential of housing as a pension. Here, one of the broad messages is that homeownership may facilitate distribution across the life cycle, but it is not generally a vehicle for distribution across income classes. The general picture for Europe is that people with the highest incomes from the labour market also tend to build up the largest

J. Doling and M. Elsinga, *Demographic Change and Housing Wealth: Homeowners, Pensions and Asset-based Welfare in Europe*, DOI 10.1007/978-94-007-4384-7_5,
© Springer Science+Business Media Dordrecht 2013

housing assets and other forms of wealth, as well as acquire the largest pension rights. So, whatever European governments might do to promote homeownership as a pension, this will not generally contribute to the large-scale reduction of the risk of poverty among older people.

5.2   Income in Kind

Income in kind or imputed rent, representing the flow of services from a house, can be considered in both gross and net terms, the difference between them being the costs incurred in consuming the house. For homeowners, the amount of the net income from their home will be affected by taxation arrangements. Thus, in countries where tax is incurred on the notional or imputed income derived from their home, the net income will be reduced. More frequently, however, the greatest reduction of the occupier's income in kind received from housing will be the cost of loan repayments. Thus, it can be expected that in countries, where there are developed mortgage markets and total housing debt is equivalent to a high proportion of GDP, total housing income in kind would be considerably reduced below its full gross amount. Data presented in Chap. 2 (especially Fig. 2.6), however, indicate that most owners over retirement age will have paid off all or at least most of their housing loans to become outright owners. The assumption that this is the case for the target group here allows a simplification of the procedure for estimating income in kind.

In fact, several different methods can be identified (Frick and Grabka 2002). These include the market value approach, which is based on the level of real rents paid for comparable properties less any part of those rents that might be related to heating or utilities, and the capital market approach, which takes the market value of the home arguing that the imputed rent should be defined as a rate of return on that value. With both approaches, there are specific methodological issues. The market value method presents problems of identifying market values for comparable properties, for example, in countries where the rental market is small or subject to rent control, or where the rental and homeownership dwellings tend to be different, perhaps the former being single family dwellings and the latter apartments, for example. With the capital value method, the decisions about the appropriate rate of return will have a systematic effect on the size of the estimates.

For the purposes of providing an indication of how homeownership would contribute to retirement income across European states, our first set of estimates is based on macro-data. In this, the absence for some countries of some key variables, such as average prices, necessitates the use of approximations. Nevertheless, the approach provides results for all the EU25 member states in a form that allows some general indications about both the scale of the income in kind as well as differences between countries. Estimates for each member state have been derived by multiplying the average market price of a home-owned dwelling by a rate of return of 4%. This particular value has been taken as an approximation based, firstly, on the arguments reported by Frick and Grabka (2002) that a return of inflation plus 2% represents a reasonably safe investment and, secondly, on the generally low levels of inflation through most of the EU in 2003, this being the time base for the present exercise. Additionally, a figure of 4% recognises that owners face continuing costs of repairs and maintenance necessary in order to maintain the capital value of their home and that there may be specific taxes, for example, on imputed rental value in a few countries only (see Table 2.3). Insofar, as there is variation of rates of inflation, repairs and maintenance costs and tax liabilities across member states, the use of a single figure of 4% will necessarily lead to overestimates for some countries and underestimates for others, an outcome which is probably acceptable given the aim of producing measures that are indicative.

Likewise, because data from different sources are used, together with approximations for some countries, one consequence may be that the measures of housing income reported in Table 5.1 are over- or underestimates for some countries, so that caution in interpretation is particularly necessary. Caution is in any case appropriate since these estimates in each member state rest on an assumption that the price distribution of national stocks of owner-occupied housing is similar as between younger and older owners. In other words, there is an assumption that average house value is not related to the age of the head of household, though it may of course be systematically related to other characteristics of national populations such as income, a point to which we return later in the chapter.

In order to facilitate comparison across member states, the estimates of income are expressed as a proportion of GDP per household. Again, there is no differentiation here as between younger and older households, and GDP is not in any case the same as household income. Nevertheless, as a measure of the hypothetical distribution equally across its households of the total income produced in each member state, GDP may be considered to be a common yardstick.

Column 1 of Table 5.1 provides measures of the relative scale of the contribution to the average household in each member state. With a median value of 12.05%, in some (Luxembourg and Latvia, notably), income in kind appears relatively modest at around 5%, while in others (Portugal and Greece of the older member states, and Lithuania, Poland, Slovakia and Slovenia of the newer ones), it may be over 20%. On these estimates, the relative contribution of housing income in kind appears particularly large in the Eastern and Mediterranean countries compared with those in other regime groups.

An additional view of the income in kind enjoyed by European homeowners is provided by the micro-data from EU-SILC. Being based on owner occupiers aged at least 65 years, this better fits the issues considered in the book as a whole, but here too there are a number of characteristics of the data that limit the conclusions that can be drawn. In particular, the definitions of imputed rent have been provided for each country by its own statistical office so that they do not necessarily match each other exactly, and it is a measure that is not provided for all countries anyway. Further, and in contrast to the estimates in Table 5.1, those in Table 5.2 are not decreased in relation to outstanding mortgages.

Table 5.1 Housing income (macro-data)

	Average income in kind as % of GDP per household	Average income in cash as % of GDP per household	Average total income in as % of GDP per household
	(1)	(2)	(3)
Corporatist			
France	10.9	13.6	24.5
Germany	8.4	10.4	18.8
Austria	9.3	11.7	21.0
Belgium	6.3	7.9	14.2
Luxembourg	4.8	5.9	10.7
Netherlands	12.2	15.3	27.5
Social democratic			
Denmark	8.5	10.7	19.2
Finland	7.9	9.9	17.8
Sweden	10.0	12.5	22.5
Mediterranean			
Greece	22.6	28.3	50.9
Italy	9.4	11.7	21.1
Portugal	24.7	30.9	55.6
Spain	17.2	21.5	38.8
Cyprus	9.2	11.4	20.6
Malta	10.5	13.1	23.6
Liberal			
UK	14.2	17.8	32.0
Ireland	11.9	14.8	26.7
Eastern			
Czech Republic	17.9	22.4	40.3
Estonia	13.5	16.9	30.4
Hungary	14.6	18.3	32.9
Latvia	3.9	4.8	8.7
Lithuania	28.3	35.3	63.6
Poland	24.3	30.4	54.8
Slovakia	27.3	34.1	61.4
Slovenia	25.7	32.1	57.8

The pattern with respect to regime groups is not as clear as in Table 5.1, but the estimates are similarly high for the Mediterranean countries. Overall, however, Table 5.2 estimates are higher than Table 5.1 estimates. The median value of the estimates for income in kind for the countries included in Table 5.2 using the macro-data is 11.4 (which is close to the median of 12.05 for all 25 member states) and using the micro-data is 25.5. Clearly, then, the micro-based estimates are higher, by a factor of about two. The information available does not allow definite conclusions about why this is so. Indeed, since Table 5.2 gives net estimates, it might be expected that they would be lower, rather than higher. To the possibility of systematic error in the macro-data anyway, however, there may be added differences between GDP

Table 5.2 Composition of total income of owner occupiers over 65 years old 2006 (%) (micro-data)

	Income from work	Pension	State benefits	Income from financial assets	Imputed rent
Corporatist					
Belgium	1	65	2	3	30
France	2	71	4	4	20
Social democratic					
Denmark	5	67	8	5	14
Finland	4	75	3	3	16
Sweden	5	67	1	3	24
Liberal					
Ireland	8	63	2	1	26
Mediterranean					
Italy	6	65	1	1	27
Portugal	5	56	10	1	27
Spain	5	67	2	1	26
Eastern					
Hungary	4	71	3	0	23
Slovenia	4	49	24	1	21

Source: EU-SILC (2006)

(Table 5.1) and personal income (Table 5.2), with the former being routinely higher than the latter. It is also possible that the average value of homes owned by older people (Table 5.2) is higher than the value of those owned by the population in general (Table 5.1). Whereas it cannot be concluded which set of figures are the more accurate, it can be said that the lower estimates are not trivial in size and the higher ones are not out of line with the levels usually applied in international studies of housing cost to income ratios (see, e.g. Stone 2006).

5.3   Income in Cash

The income in cash embedded in a home can also be calculated from its market or capital value. A house, whether it is sold to another homeowner or acts as the basis of a financial product, would provide the owner with either an income directly or a lump sum, which could be converted through an annuity into an income. The amount of the income would be reduced as a result of any outstanding loans taken out in order to purchase the home. In other words, the net capital value and net income in cash would be proportional to the current proportion of the equity held by the owner. As with the estimates of income in kind, the assumption is made that older homeowners are typically outright owners. Consequently for the present exercise, a simplifying assumption is that the relevant amount is related to the full market value of the home.

Here, income in cash is calculated by using an arbitrary value of 5% of net equity. This is not strictly how the value of a reverse mortgage or other equity release financial products would be calculated, because these are likely, since it will influence the length of time before repayment will be made, to factor in the present age of the householder. Simply, the older the homeowner and the shorter their life expectancy, the higher the income that could be generated by a given amount of capital. Also taken into account will be expected future house price inflation and interest rates. However, assuming a product taken out at 60–65 years and a life expectancy in many member states of maybe 20 years beyond this, as well as general inflation in the early 2000s, 5% may be a reasonable approximation. These estimates are shown in column 2 of Table 5.1.

Income in cash has also been estimated using EU-SILC. Although this source does not provide a direct measure of house values, this has been estimated by working back from the imputed rental value, which is given, assuming that this will be approximately 5% of capital value. From this, two models and four measures have been estimated:

- Model 1: Sell the house, pay the proceeds into a deposit at a per cent interest rate and run this value down over a period of either 15 or 20 years. This means the household will have to move to a rental dwelling: the proceeds from the released equity are consequently corrected for the monthly price of the 'average' rental house. In Table 4.2, this is strategy 6.
- Model 2: Take a reverse mortgage against 75% of the value of the home over a period of either 15 or 20 years at a 6% interest rate. In Table 4.2, this is strategy 5.

In these models, the income from equity release is expressed as a percentage of the disposable household income. The house price levels used are from 2008, the peak of house prices as became clear in the last years. However, the general patterns over countries should remain valid. The results are shown in Fig. 5.1. Although there are differences in detail compared with the estimates in column 2 of Table 5.2, overall the scale of the contribution is at least of the same order of magnitude. More so than with the macro- and micro-estimates of income in kind, here there is some similarity.

There are a number of important variations across the estimates. Firstly, the obvious conclusion is that using the house equity over 15 years produces an income in cash higher than if using it over 20 years. Secondly, and less obviously, income in cash from an outright sale even accounting for the need to pay rent in cash for somewhere else to live is higher than the reverse mortgage strategy. Thirdly, accommodating the different scales, the income in cash for both strategies is generally higher for households in Denmark, Spain, Ireland and Hungary. Each of these countries falls in a different welfare regime, so there is no clear pattern that reflects the welfare regime groupings.

5.4   The Adequacy Criteria

Once the amount of these two sources of housing income potentially available to European homeowners is estimated, the second stage of examining how homeownership may function as a pension consists of identifying appropriate criteria against which they can be assessed. From a range of sources, including those related to the EU (European Commission 2006), the World Bank (Holzmann and Hinz 2005) and the OECD (Disney and Johnson 2001), it is possible to identify the primary goal of pension systems as being adequacy in the senses of firstly maintaining the pre-retirement standard of living of European citizens and secondly protecting them from the risk of poverty. The aim here, then, is to assess the extent to which the use of the income that might be derived from housing would contribute to meeting these two criteria.

5.4.1   Maintaining Living Standards

5.4.1.1   Income in Kind

Starting with the estimates in Table 5.1, on the face of it, the size of the estimated incomes in kind indicates that the person who, at the point of their retirement is an outright owner of the home they occupy, is able to go on living in that home, and in that way their standard of living – or at least its housing element – is maintained. This is, of course, consistent with Castles' thesis that homeownership is a means of distribution over the life cycle, that by the time they reach retirement age homeowners are typically outright owners and can consequently get by on a smaller pension. As indicated earlier, in comparison with those of Table 5.1, the estimates of imputed rent provided by Table 5.2 are generally higher as a proportion of income, accounting very approximately, over the member states included, for a quarter of the total income. Irrespective of whether the lower or higher estimates are used, however, they indicate that income in kind from housing makes a considerable contribution to maintaining the living standards of older European homeowners.

5.4.1.2   Income in Cash

The housing income that owners could derive from the equity in their home provides them with a clear advantage over renters. Column 2 of Table 5.1 indicates that this might contribute amounts from just under 5% to more than 30% of an average household's share of GDP. As indicated, these are similar in scale to the estimates in Fig. 5.1. Both can be put alongside the figures in column 1 of Table 5.3 which shows the average incomes of those aged over 65 relative to those under 65. In many

Table 5.3 Pension and income characteristics

	Relative median equivalent income of age 65< compared to 64> (%)	Poverty rate (%) 65+ years	Poverty rate (%) 0–64 years
	(1)	(2)	(3)
Corporatist			
France	90	16	13
Germany	88	16	15
Austria	93	17	12
Belgium	76	21	14
Luxembourg	101	6	12
Netherlands	84	7	13
Social democratic			
Denmark	71	17	10
Finland	75	17	10
Sweden	77	14	11
Mediterranean			
Greece	78	28	18
Italy	95	16	20
Portugal	76	29	19
Spain	77	30	18
Cyprus	–	52	10
Malta	90	20	14
Liberal			
UK	74	24	17
Ireland	62	40	19
Eastern			
Czech Republic	83	4	9
Estonia	76	17	19
Hungary	87	10	12
Latvia	80	14	17
Lithuania	89	12	15
Poland	113	6	18
Slovakia	89	12	22
Slovenia	87	19	9

Source: European Commission (2006)

member states, older people enjoy average incomes only slightly lower than younger people, that is, at least 80%, and in two cases, Luxembourg and Poland, is over 100%. The addition of housing income in kind and in cash at levels suggested by Table 5.1 (column 3), Table 5.2 and Fig. 5.1 indicates that the combined housing and non-housing income, which in the case of most countries is in excess of 20% of GDP, might result in the average older person actually being better off than the average younger person. If, further, the lack of housing loan repayments for this age group is factored in, overall, these estimates suggest that if older homeowners were to routinely convert their housing equity into income, on average they would be able to consume at levels similar to if not greater than they did when younger.

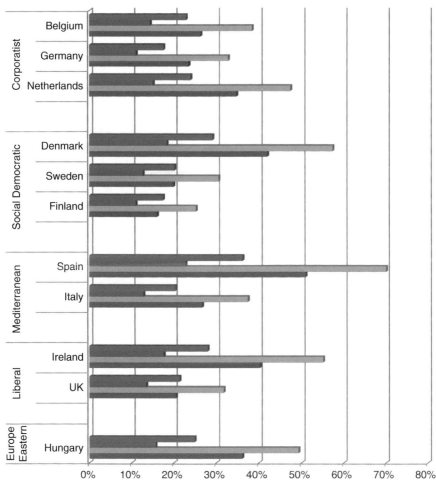

Fig. 5.1 Estimated additional income from housing assets as a percentage of disposable household income of older people (65+) (micro-data) (Source: EU-SILC 2008)

A note of caution is appropriate in that the figures in Table 5.3 do not distinguish between buyers and renters. The argument nevertheless remains that housing equity potentially contributes to both the non-housing as well as the housing elements of the standard of living of older homeowners. It may not in itself provide support for older renters, and in a relative sense, may even make them worse off, but there seems little doubt that housing income in kind and in cash would provide older

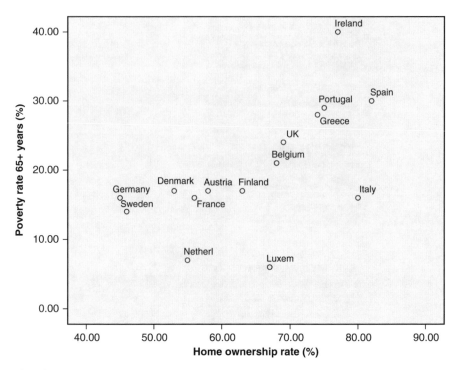

Fig. 5.2 Homeownership and poverty rates

homeowners with the means of considerably enhancing the income they get from other sources. Moreover, this conclusion holds whether our higher or lower estimates are used.

5.4.2 *Reducing the Risk of Poverty*

The estimation of the potential impact of housing income in kind and in cash on poverty reduction is less clear. One view is provided through the significant correlation between poverty rates (Table 5.3 column 2) among older people and homeownership rates. Those older member states with the highest poverty rates also tend to have the highest homeownership rates. This appears consistent with Castles' argument that the comparison of welfare systems *between* the Old World (western Europe) and the New World (Australia, New Zealand and Canada) needs to recognise that low levels of benefits through welfare systems in the latter have in practice been compensated for by high benefits through the ownership of housing (Castles 1998a, b).

Figure 5.2, covering the older member states, suggests that the comparison might also apply *within* the Old World: the greater poverty rates in some member states are compensated by higher rates of homeownership. Specifically, the scatter plot suggests a group of countries, the southern or Mediterranean states plus the Liberal countries of Ireland and the UK, having high rates of both old age poverty and homeownership. Such a conclusion has also been found in an analysis of nine of the older member states which demonstrated that housing income – in kind and in cash – had the greatest impact on reducing the incidence of poverty among older people in those states with most homeowners (Lefebure et al. 2006). At a macro-level, therefore, homeownership appears to offer a potential solution to high rates of poverty among older Europeans.

Such a conclusion is also supported by estimates using the micro-data of EU-SILC. Presented in Fig. 5.3a, b, these are based on the same two models as presented in Fig. 5.1: model 1, sell and rent, and model 2, take a reverse mortgage, each for 15- and 20-year periods. For all, income from equity release is expressed as a percentage of the disposable household income. The results indicate the potential impact on households living on incomes that otherwise place them below or above the poverty line in their country. In general, those living below the poverty line would receive a greater proportional boost to their income, irrespective of whether they pursue the sale and rent or the reverse mortgage strategy. In most countries, for those in poverty, housing income, relative to actual income, is twice as high as it is for the non-poor, the main exceptions being Finland, Belgium and Hungary.

Whereas these results point to the potential of housing income in reducing the risk of poverty, there are (at least) two critical issues. Firstly, the evidence in Chap. 2 indicates that in general those who buy their homes tend to have higher incomes than those who rent. Because people who have had low incomes during the working periods of their lives will also tend to have low pensions and low levels of savings, the risk of poverty will generally be higher among renters than among owners. It follows that those older Europeans who rent their homes and are in poverty cannot access housing income in kind or in cash that would improve their income position. Consequently, there is no potential in housing equity to overcome the poverty of many older Europeans.

Secondly, even confining consideration to homeowners, it is relevant to consider whether the absolute size of the gain from housing income is higher for those in poverty than those not in poverty. In other words, would the addition of housing income reduce the spread or distribution of incomes (housing and non-housing combined), or would it tend to improve the position of those at the bottom a little, and those at the top a lot? Or, do older people with the lowest cash incomes from pension and other sources tend to have a lot of housing equity which could compensate for their low cash income?

Measures of both housing and non-housing income derived from EU-SILC allow an assessment for each member state individually. The correlation coefficients between the non-housing income of homeowners aged over 65 and the imputed rent they enjoy are, with the exception of just one member state, positive and significant. For the exception, the Netherlands, the correlation is positive but not significantly different from zero. The general pattern over the EU member states, then, is clear: in each country, older people with the smallest incomes tend to have the least housing assets.

This pattern is further confirmed by Table 5.4 which uses the same variables (housing and non-housing income) in order to indicate the proportions of older homeowners who could be classified as asset-rich, cash-poor. Column 1 indicates the proportion who have incomes below the median level, for all older owners, and

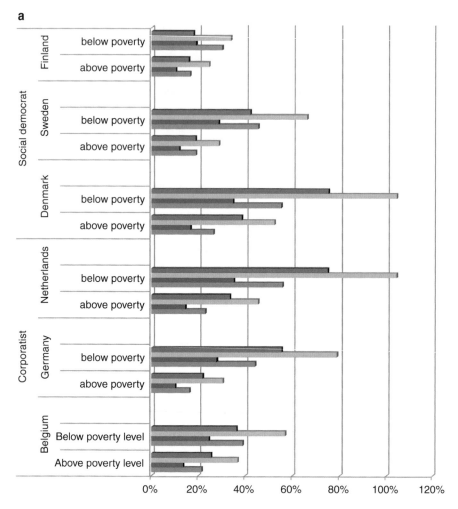

Fig. 5.3 (a) Estimated additional income from housing assets as a percentage of disposable household income of older people (65+) (b) Estimated additional income from housing assets as a percentage of disposable household income of older people (65+) (Source: SILC 2008 and authors' calculations)

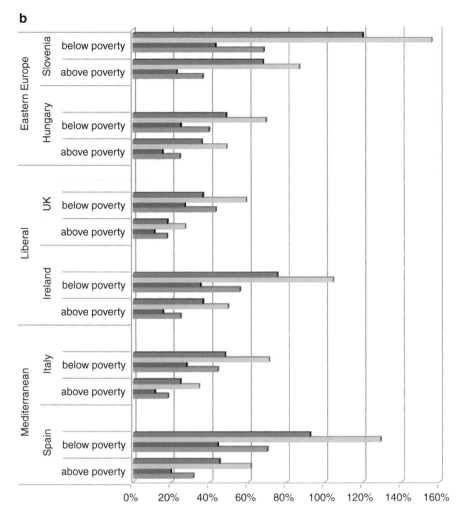

Fig. 5.3 (continued)

housing assets above the median level for all older owners. If older owners were equally distributed across income and asset categories, 25% of older owners would fall into this group. Column 2 indicates the proportion who lie in the upper half of the housing asset distribution, but the lowest quartile of the income distribution. Labelled as asset-rich and cash very poor, equal distributions would mean that 12.5% of older homeowners fell into this group.

Table 5.4 Asset and income groups: homeowners aged over 65

	Asset-rich; cash-poor (%)	Asset-rich; cash very poor (%)
	(1)	(2)
Corporatist		
Austria	18.6	8.5
Belgium	14.8	5.6
Luxembourg	20.2	8.8
Netherlands	27.3	15.6
Social democratic		
Denmark	23.4	11.0
Finland	17.9	8.3
Sweden	20.8	8.4
Mediterranean		
Greece	17.2	5.9
Italy	17.5	7.6
Portugal	15.8	6.0
Spain	19.4	8.7
Liberal		
UK	20.1	9.5
Ireland	22.3	10.3
Eastern		
Cyprus	12.0	5.3
Czech Republic	20.1	8.4
Estonia	20.4	9.2
Hungary	19.0	6.6
Latvia	21.9	9.1
Lithuania	22.6	8.4
Poland	18.4	7.5
Slovenia	19.7	5.9

Source: European Commission (2006)

Again with the exception of the Netherlands, for all countries, less than 25% of owners could be classified as asset-rich, cash-poor, and less than 12.5% asset-rich, cash very poor. This is consistent with the simple correlations. It indicates that while some of those older homeowners who have the lowest incomes, and in that sense are most in need of a boost to their incomes, have considerable housing assets which could in principle provide that boost, considerable housing income is more likely to be a characteristic of those who already have higher incomes. Here, too there are country differences. The Mediterranean countries – Greece, Italy Spain and Portugal, but also Slovenia and Cyprus – have the lowest proportions in the asset-rich, cash-poor and very poor groups. This suggests that in those countries in particular, housing income has the least potential for overcoming poverty in old age. Table 5.4 thus might be taken to counteract the apparent message of Fig. 5.2: high homeownership rates do not necessarily mean that homeownership can ameliorate old age poverty.

These results also indicate that the potential contribution of housing towards income in old age is likely to be different in different member states according to differences in both pension and housing systems, but also to the distribution within their populations of housing, pension and other assets. The stylised picture of older people being poor or at least at greater risk of poverty, perhaps because of pension provision that is at a low level relative to incomes from work, with housing wealth providing a compensating mechanism, is not necessarily accurate. Indeed, on the basis of this evidence, it seems probable that homeowners are generally likely to have higher incomes than non-owners. In addition, those with the highest cash incomes could also receive the highest income from housing, perhaps increasing rather than reducing income inequalities and not necessarily making any great reduction in the risk of poverty.

5.5 Sustainability

The second of the two criteria used in this chapter is that pension systems require financial soundness both now and in future. This means that the costs involved in ensuring a given level of income in old age are not subjected to increases, or the reverse that, while keeping costs level, the given level of income will not be decreased. Of particular concern in many evaluations is the change in dependency ratios which may have particular implications for pay-as-you-go pension systems since, at any one time, they require those currently working to fund those currently retired. In the case of homeownership, whereas in the past its investment potential may have made it an attractive option to many European citizens, and thus contributed to the growth of homeownership sectors, as argued in Chap. 2, growth has also frequently been enhanced by state policies. These include both demand-side subsidies such as tax breaks of various types to homeowners and supply-side initiatives, primarily the privatisation of formerly social rental housing. The sustainability issue, therefore, concerns the nature of these policies and potential threats to their continuation that might affect the potential contribution that housing assets might make to the income needs of older people.

5.5.1 Demand-Side Subsidies

The European Central Bank has concluded that member state governments have set in place a number of interventions that 'are directed explicitly at promoting home-ownership, in many cases explicitly for low-income households' (ECB 2003: 35). As indicated in Chap. 2, these interventions include tax breaks on housing-related investments, such as relief from tax on imputed rental value of owner-occupied housing and reductions on loan interest. The future of such policies will in part be contingent on the desire and willingness of member state governments further to promote homeownership. Such desire may arise for reasons of their potential contribution to resolving the 'pension crisis'. Those governments that couch the problem in these terms could take the view that, relative to maintaining the generosity of present state pension schemes, tax subsidies to encourage homeownership present a fiscally cheaper alternative.

The future will also be contingent, however, on the fiscal difficulties facing European governments. These may be prompted by concerns about maintaining the growth and stability conditions on which inclusion in the eurozone is dependent. In the Netherlands, interest rate subsidies have been so large as to provoke a challenge from the European Union (European Commission 2006) and recently from OECD (2011). Cutting the level of tax payments set against housing costs would enable governments to reduce tax burdens overall or to reduce public debt. But the fiscal difficulties may also be a consequence of the changing dependency ratios that are the basis of the pension crisis and may lead to governments seeking other tax savings in order to protect pensions.

5.5.2 Privatisation

Also as discussed in Chap. 2, in many member states, homeownership has been fuelled by the privatisation of social housing stock. In fact, even if the numbers involved are not always large, most EU member states now allow the sale of social housing (Norris and Shiels 2004), while in some of the newer member states, the sale of former state-owned housing has often taken place on a massive scale (UNECE 2006). While the overall effect of these policies has been to boost the size of homeownership sectors, and in the case of some countries such as the UK and some of the former communist countries has been considerable, the long-term impact may be limited. Firstly, the addition to the supply of dwellings available to homeowners may have a downward influence on other sources of supply so that the net increase may be reduced. Secondly, the finite size of social housing sectors and the speed of privatisation will set limits to the extent and prolongation of the boost.

5.6 Conclusions

So, how would homeownership perform as a pension? The analyses reported in this chapter have provided evidence of how the income embedded in housing might perform in relation to the two dimensions of the adequacy criterion – maintaining living standards and reducing the risk of poverty. There is not a simple answer that applies uniformly across the member states; nevertheless, housing income in kind may enable older homeowners to live rent-free and thereby continue their former standard of housing. Generally, the estimates presented indicate that housing income, both in kind and in cash, do and could make a considerable contribution to maintaining the former standard of living of older people. Income in kind alone appears to constitute, on average across member states and households, about a quarter of total income. It is clear that if to this were added an income in cash created by realising the full equity of the home, the boost to existing incomes would indeed be considerable. It follows that, all other things being equal, the release of housing assets in a way consistent with the life cycle model, would result in the average older European enjoying a higher standard of living. Moreover, the option of selling and moving to a rental dwelling seems to be better value for money than buying equity release products.

But all other things are not necessarily equal, not least because our analysis has implicitly made an assumption about the continuation of existing pension systems. The conclusions would undoubtedly be affected by a further consideration: should the question of the performance of housing as a pension be based on the principle of housing as a complement to, or of housing as a substitute for, tax-funded pension systems. The use of housing assets to support the consumption needs of older households is, on the one hand, an option for households providing them with an opportunity to complement other forms of saving, including through state pensions. As such, the opportunities for a financially comfortable, high-consuming old age are considerably enhanced. In these circumstances, housing wealth is a *complement* to other income. On the other hand, housing assets also present an option to governments wishing to find substitutes for pension and other public responsibilities for meeting the costs of old age. Encouraging households to rely less on the state by saving more, through housing and other forms of investment, could seem an attractive way of combating the fiscal problems of funding existing state pension systems. In these circumstances, housing wealth becomes a *substitute* for taxation.

Whatever the outcome for maintaining former living standards, the potential effect of housing income on reducing the risk of poverty is also important. On the face of it, the potential would appear great. Member states with high rates of poverty among older people also have high homeownership rates, suggesting that income from the latter would turn asset-rich income-poor households into income-rich ones. However, the indication of our estimates is that in general there is a positive correlation between those with most housing assets and those with most cash income anyway. In other words, the potential of housing equity would be greatest for those who, on objective grounds, might be deemed to need it least. Generally, housing assets would not appear to be a mechanism for reducing inequality across populations; they may help to smooth income across the life cycle but not necessarily across income groups; they constitute a means of horizontal not vertical redistribution.

There is a further consideration. Our estimates, and indeed much of the discussion about the use of housing assets, focus on the growing majority of homeowners. But of course even if two-thirds of Europeans are homeowners, by definition, one third are not. For them, the potential of housing equity is at best an irrelevance: there might be no direct impact on them if housing wealth is accessed as a complement. At worst, if it is used as a substitute, as part of a general reform of pension entitlements, the critical issue would be how their pension rights were affected. Moreover, given doubts about the future sustainability of homeownership sectors at their present size, it is possible that the number of non-homeowners may actually grow.

Chapter 6
Lessons from East Asia

6.1 Introduction

This chapter is the second of two that respond to the third of our general questions about how housing assets, if systematically used, might perform as a pension. The specific objective is to identify and evaluate the lessons to be learned from international experience. A number of societies beyond Europe have long been reliant on homeownership as a supplement for, or even the basis of, pension income. Specifically, the focus here is on developed East Asian societies – Japan, Korea and Singapore – which have reached, or even surpassed, the economic status of some European societies. Whereas elements of their approach to welfare bear some similarity to that in Europe, there are also important differences of emphasis. They have been resistant to the building up of welfare states and citizenship rights, placing considerable emphasis on the responsibilities of the family. Supporting this, housing policy and advancing homeownership have played an important part in both the economic success of these countries and in supporting universal living standards. In some respects, then, they have been forerunners of property asset-based welfare approaches, the experience of which may inform European thinking about different welfare futures.

Moreover, from the perspective of a Europe struggling to find a way forward from the economic crisis of the late 2000s, East Asian policy developments after the Asian financial crisis, which took place more than a decade ago, are illuminating. This had a large impact on currency and stock market values as well as on housing markets. As the East Asian region entered a more volatile economic era in which high-speed economic growth and full employment was no longer assured, the prevailing housing model came into question. Housing markets began to feature large numbers of homeowners with negative equity and increasing economic inequality between different tenures, types of property and cohorts of renters and owners. In the 2000s, in the light of incongruities between housing market conditions and welfare demands, many East Asian governments commonly adjusted

J. Doling and M. Elsinga, *Demographic Change and Housing Wealth: Homeowners,* 119
Pensions and Asset-based Welfare in Europe, DOI 10.1007/978-94-007-4384-7_6,
© Springer Science+Business Media Dordrecht 2013

housing market regulations, the focus of policy on home purchase and dependency on homeownership as a means to meet welfare needs. A particular salient development has been the emergence of public equity release schemes that explicitly target housing wealth as a means of supporting incomes in old age.

Both the general and specific, post-crisis, lessons can be examined against the two criteria, the adequacy and sustainability tests, used in Chap. 5. The broad message is that homeownership can work relatively well as a pension in a context in which the wider welfare system is not based on a strong commitment to vertical redistribution. In such contexts, pension systems and ownership mutually reinforce. Nonetheless, this approach to welfare has its limitations, and in recent years, the pitfalls of dependence on homeownership have become apparent. With different emphases, these economies have responded with two broad strategies. The first has been to try to improve access to housing equity – both building up equity and realising it – as well as the stability of the housing market, thus reinforcing the importance of home-ownership in the pension equation. The second has been to introduce social protection measures that compensate for the uneven distribution, performance and access to housing assets and the lack of public welfare alternatives. There has thus been both a reinforcement of the existing approach based on homeownership and, at the same time, through a greater state role in redistribution, a diminution of its position as a pillar of household security.

In the next section of the chapter, the broad approach to welfare and housing in the more advanced economies of East Asia is presented as a background to the country case studies that follow it. The subsequent three sections introduce the links between housing and pension systems in each of Singapore, Japan and South Korea in turn. Finally, by considering their experiences against the adequacy and sustainability criteria, we identify a number of lessons that might apply to Europe.

6.2 Housing and Welfare in East Asian Societies

Although the notion of welfare regimes has provided insights into European welfare states, explaining the dynamics of welfare provision in recently industrialised East Asian societies has proved more problematic. East Asian countries experienced rapid economic growth in the latter decades of the last century but maintained both minimalist approaches to social welfare and relatively stable political hegemonies. Early analyses focused on sets of social and cultural values associated with Confucianism (see Jones 1993). However, considering local power practices based on a model of the 'developmental state' (see Johnson 1982) has provided better insights. Developmental states feature a brand of economic nationalism within which corporate and bureaucratic elites form alliances aimed at driving economic growth. This form of governance has been associated with 'productivist' welfare regimes in which the target of economic growth is predominant (Holliday 2000; White and Goodman 1998).

There are several important features of this model of welfare. Firstly, the predominance of economic objectives is truly such that social policies have been subordinate. There has not been the development to a significant level, as has been common in Europe, of social or citizenship rights taking form through social protection measures. While the major task for governments has been to set the path for growth and employment opportunities, the associated role of their populations has been to work hard and through that protect themselves. Secondly, and supporting the individual, has been a reliance on the family as the basic unit of society through which individuals meet the welfare needs of family members. In reality, families commonly cover gaps in state provision by providing informal social insurance for their members, especially in old age.

Thirdly, and supporting the first two, has been the role of homeownership (see Groves et al. 2007; Ronald 2007). The promotion of homeownership has been strongly implicated in rapid economic growth in this region in the later decades of the twentieth century. On the one hand, expanding owner-occupied housing sectors drove intensive economic growth and urban development. On the other, increases in homeowner households and house prices supported the asset base of family-orientated welfare practices, offsetting the need to build up an onerous and costly welfare state and facilitating the focus of government resources on industrial expansion. Further, homeownership has provided the basis of care for family members in terms of shelter, an income in kind and a financial reserve to draw upon in case of hardship. Housing not only provides the financial basis for retirement, it also features strongly in reciprocal exchanges between generations where care in old age is often provided on the basis of cohabitation and/or inheritance. This approach to housing provision has arguably supported substantial levels of security and social stability with limited recourse to government build-up of expensive welfare services.

In recent decades, however, particularly since the Asian financial crisis of 1997, this type of asset-based welfare in East Asia has been in retreat, or at least the role of housing assets in the welfare system has been adjusted with the apparent end of an era of smooth economic growth in which house prices only improved (Ronald and Chiu 2010; Ronald and Doling 2010). These economies have also faced significant demographic transformations, undermining family-based welfare practices and fragmenting households and life courses. Additionally, there have been transformations in political conditions that have undermined the dynamics of developmentalism and productivism, with a growth in public spending levels on the one hand and economic liberalisation of markets on the other (Kasza 2006; Peng and Wong 2004; Weiss 2003). The impact of the global financial crisis is not yet clear, and while economic output and housing markets initially crashed, there was very rapid recovery in 2009. Nonetheless, it has arguably reinforced the sense of vulnerability and economic volatility that has shaped housing and welfare policy reform in the last decade. The role of housing and pension policy and the function of the family and family wealth in welfare are ostensibly undergoing transformations in these national contexts. The trend has been towards diversification across policy systems and greater application of public insurance measures. Homeownership policy and housing property wealth, nevertheless, continue to shape these transformations.

Whereas this describes some common features of the East Asian model, the particulars vary from country to country. Considering the development of housing systems and how asset-based welfare has manifested itself in Singapore, Japan and South Korea, however, provides a basis for identifying common lessons that inform debates in European countries.

6.3 Singapore

Singapore has not been regarded as having a 'welfare state', and needy Singaporeans can often get help only through a handful of financial assistance schemes based on providing short-term relief (Yap 2002). In general, people are expected to rely on their own savings and assets, the accumulation of which the government has sought to support through a combination of compulsory payments (taken from monthly salaries with a matched copayment from the employer) into an individual account in a Central Provident Fund (CPF). Savings built up in the provident fund can be kept in a pension fund and accessed in later life but are normally transferred earlier into housing with a residual retained in order to pay a lump sum on retirement and provide a small retirement pension income.

Over the last 50 years, the system has been developed in order to keep pace with socio-economic transformations as well as developments in the housing market that have impacted on the effectiveness of housing investment as an alternative or supplement to retirement income. Essentially, the embeddedness of the homeownership pillar of Singaporean welfare has meant that the state has had to persist with a mass, subsidised homeownership programme. Nonetheless, it has also sought to diversify the asset pillar of welfare by containing the amount of pension funds that can be invested in housing and placed greater emphasis on non-housing investment vehicles.

6.3.1 The Housing System

The ability of members to transfer their CPF savings into housing properties has meant that the vast majority of people have been able to buy their own homes to live in, while these have also functioned as investment products that augmented in value over their lifetimes. It has been imperative therefore that a system of homeownership be established that provides access to affordable housing properties and that house price increases are maintained.

Homeownership rates in Singapore expanded from 29% in 1970 to 92% by 2003. There have been a number of necessary conditions by which the government has been able to control this growth. The primary institution is the Housing Development Board (HDB). Established in 1961, it produced the first basic high-rise flats and later began to encourage eligible households to purchase 99-year leases and thus become owner-occupiers. In 1971, the resale of public-leasehold housing units was permitted, creating a secondary market of owner-occupied housing, where the state continued to regulate eligibility and subsidy. What is unusual about this system is that it constitutes a public form of homeownership with the state retaining a central role in the housing market, controlling the flow of stock, the price of new units and who is eligible to buy at subsidised rates. This contrasts starkly with European programmes to expand homeownership which have focused on the deregulation of housing provision and mortgage systems, the transfer of tenure from rental to owner-occupied forms and an emphasis on the private sector.

Following the 1966 Land Acquisition Act, the state began buying up land from private owners at below market rates, which allowed cheaper provision of public housing development. By the early 1990s, the state owned around 90% of the total land supply. The system of selling leases to properties on state-owned land has facilitated accelerated system expansion, and as with each cycle of construction, the state has been able to recover costs in order to finance the next cycle (Chua 2003). The HDB has not only provided dwelling units but also mortgages and mortgage insurance for purchasers of new and used public owner-occupied flats.

6.3.2 Cross-Funding Housing and Pensions

Since 1955, the vast majority of Singaporean workers have contributed a monthly proportion of their wages to the CPF. Membership of the scheme stood at 3.02 million in 2005/2006 with total balances at US$68.5 billion. On reaching 55, members can withdraw the majority of their savings once a minimum has been put aside, which is later released as monthly payments once they reach 62. Monthly contributions are allocated to three accounts, which can be used for different purposes. For example, the special account is reserved for old age and/or investment in retirement-related financial products, while the ordinary account can be invested into CPF-approved investment vehicles and, more importantly, be used to buy a home.

Investments in HDB owner-occupied homes thus facilitate the transfer of credit built-up in the compulsorily saving scheme, the primary source of social insurance, into a 'housing account' with transfers allowed for both down payments and mortgage repayments. CPF savings can also be used to cover survey and legal fees, charges related to the use of the property, renovation or repair. In 1981, the Approved Residential Properties Scheme allowed CPF savings also to be used for private housing.

In principle, then, the CPF is a pension fund. However, unlike European provident funds, the Singapore CPF has particularly targeted housing wealth as a pillar of pension provision. As the majority of pension saving can and normally is transferred into housing property over the lifetime, it ultimately constitutes the basis of retirement planning and security. The way CPF savings and withdrawals and HDB housing are structured is thus complementary. The idea that reduced housing costs in old age (the income in kind facilitated by owner-occupation) and the security of a large asset holding together offset a smaller pension income in retirement is largely explicit. Moreover, in 2009, the government instigated a Lease Buyback Scheme for HDB homeowners over 62 years old. Under this, the HDB will purchase any remaining years of lease, in excess of 30 years, thus giving homeowners access to equity they have built up over their housing career, while continuing to live in the property. Whereas this does not constitute a reverse mortgage product from a financial institution, it fulfils the same function (Doling and Ronald 2011).

6.3.3 Maintaining Housing Prices and Property Assets

There has been careful control of the HDB market. Over time, and in order to ensure a flow of new purchasers, restrictions on purchase in the primary and secondary market have been eased. Notably, since the 1990s, income ceilings have been raised, and single people have been allowed access to the secondary market. The government has also increasingly sought to deregulate the system in order to free up market mechanisms, but it has retained significant control to ensure the value of properties remains secure. Price falls would devalue the capital embodied in the housing stock and jeopardise the financial security of households, especially those nearing retirement.

The 1997 Asian financial crisis and subsequent economic downturn initially hit housing demand hard. Essentially, the housing ladder system ceased to support a process of house price inflation, exposing retirement savings to market fluctuations. Immediately after 1997, prices dropped by as much as 30% with specific losses strongly determined by the timing of purchase. The government halted land sales and introduced new demand subsidies for purchases by singles. Nevertheless, housing prices declined further and conditions worsened with the growing numbers of unsold units (by 2002 there were 17,500 unsold units). It was not until 2004 that the market showed signs of stable recovery.

The economic recession revealed fundamental flaws in practices of overbuilding, demand side subsidies and administered prices that were not adjusted downwards. In the 2000s, the HDB reduced output in line with the fall in demand. There was also considerable deregulation, although the HDB and CPF maintain the stability of housing assets. Prices of new flats are pegged to average household income levels to ensure that 90% of all households can afford 70% of a new, four-room flat (Phang 2007). In 2002, caps were placed on CPF withdrawals for housing to reduce risks of over-concentration in terms of asset portfolios.

The 1997 crisis also revealed the exposure of many households to market risks and the limitations of housing assets as security in retirement. Older households have, on average, around 75% of their retirement wealth in housing (McCarthy et al. 2002). Lim (2001) projected that in Singapore, 60–70% of 50–55-year-old cohort would not have sufficient funds to meet government-stipulated minimums for retirement. As much as the CPF and HBD contribute to a level of stability in the housing system, over-concentration of savings in the housing stock make the pension and welfare system vulnerable to changing socio-economic conditions.

In 2001, the Supplementary Retirement Scheme (SRS) was introduced illustrating growing concern with the potential inadequacy of CPF-funded housing investment as a pillar of old-age provision in the light of economic trends. Nonetheless, for most households, the purchase of a HDB home and the accumulation of housing property assets through moves up the property ladder continue to be central to saving and retirement strategies. Indeed, increases in house prices are, in light of the volume of pension savings invested in the housing stock, most significant to the economic security and relative wealth of Singaporeans in old age.

6.4   Japan

Public spending in Japan, as a proportion of GDP, is higher than in most other East Asian countries. Nonetheless, the Japanese welfare system has largely been orientated around forms of company and family welfare provision that are enhanced by the individual accumulation of owner-occupied housing assets (Hirayama 2010). Housing policy has sought to expand homeownership, and, as in Singapore, a large government infrastructure was established to advance home purchase. However, measures linking housing to welfare are more implicit and indirect. The government, consequently, has not been so bound to the maintenance of public housing subsidies and institutional structures. In the 2000s, Japan deregulated its housing system and reformed its pension system. A legacy of the previous era, however, is a dependence on the wealth built up in housing; essentially, homeownership remains fundamental to the economic security of the majority of Japanese in retirement.

Unlike Singapore, Japan has established unemployment insurance and a PAYG pension system (supplemented by occupational pension schemes) (see Shinkawa 2005). However, in the mid-1980s, following increasing demographic pressures, pension retrenchment became a government target. Japan has been an 'aged society' since the 1980s, and by 2000, more than 20% of the population was already over 65 years old (the largest proportion of elderly in the world). It is estimated that the ratio of working to retired people will reach 2:1 in the next decade or so (UN 2005). Japan also experienced prolonged recession in the 1990s, putting even greater stress on government resources. Since 2002, the government has sought to neoliberalise pension schemes and has given up on defined benefit targets.

Nevertheless, while pensions have been eroded and welfare resources have become stretched, the number of housing asset rich older people has expanded. It is estimated that around 86% of Japanese pensioners are homeowners, most of them un-mortgaged, with many owning more than one property (Hirayama 2010). The problem this has presented is that while many older people are housing asset rich, this has not consistently translated into being welfare rich. The state has subsequently implemented a large long-term care insurance programme. This shift in the logic of welfare provision, with a growing emphasis on the state, has also coincided with the rolling back of housing subsidies for home buyers.

6.4.1   The Housing System

The pillars of housing policy introduced in the 1950s aimed to redirect families up an owner-occupied housing ladder (Hirayama 2003). The Government Housing Loan Corporation (GHLC), in particular, provided funds for long-term, fixed low-interest mortgages, facilitating a rapid increase in urban homeownership rates from around 25% in 1940 to over 64% by the mid-1960s. Essentially, the combination of homeownership and a growing middle class was regarded as a key factor in stabilising the socio-economic conditions necessary for economic growth (Hirayama and Ronald 2007).

Government support for homeownership in Japan has been supplemented by the 'enterprise society'. In this model, the company acts as a form of family for employees, rewarding their loyalty with lifelong-employment security, age-based wage increases and a raft of welfare goods and services including housing (see Sato 2007). Company rental housing provided a means for new families to establish themselves outside the parental home while saving up to purchase their own home. Most large companies also provided subsidised housing loans for employees, which supplement the portfolio of borrowing necessary for a family to get into owner-occupation.

In the 1950s and 1960s, socio-economic conditions and GHLC loans (typically in combination with company and family loans) made owner-occupation accessible to large swathes of the working population. However, high demand and a scarce supply of urban land pushed rapid house price inflation. Increases were interrupted in 1973 by the oil crisis, and the state looked to the housing system as a means to regalvanise the economy. Policies to stimulate housing construction and purchase became more aggressive over the next 10 years, and the GHLC initiated a series of programmes to maintain access to, and growth in, the owner-occupied sector: the Step Repayment System (1979), in which repayments were lowered for the first 5 years; the Two-Generation Mortgage (1980) which allowed adult children to take over their parents mortgage and extend the repayment period; and the Supplementary Loan (1985) in which additional loans were added to the main mortgage.

By the 1980s, a cycle had formed where increasing house prices demanded the improvement of lending conditions, which encouraged house acquisition that, in turn, expanded demand for owner-occupied housing, boosting housing prices. The largest spirals in property prices came in the bubble years of the mid-1980s, especially in metropolitan areas. Between 1980 and 1990, price-income ratios for housing in Tokyo increased from 5.0 to 8.0 for a condominium and 6.2 to 8.5 for a single-family house making home purchase and the maintenance of loan repayments increasingly challenging for many.

6.4.2   Sustaining Homeownership

Government promotion of homeownership had been particularly successful in providing the growing number of middle-class households with housing property assets. However, in the 1980s, very little was done in order to address the sustainability of the property market and access for younger households. Indeed, it has been argued that the government encouraged house price inflation that ostensibly expanded the asset wealth of existing family households (Forrest and Hirayama 2009). The start of the 1990s saw a sharp decline in stock market values followed by a collapse in housing property values: urban property values dropped by between 40% and 50%. The majority of owner-occupiers consequently experienced major capital losses on their housing assets. Japan subsequently entered a prolonged period of recession between 1990 and 2002 (and again after 2008). Although housing asset stability had been considered central to the security of households, the government prioritised overall economic recovery and subsequently attempted to stimulate construction. However, despite early post-war shortages, vacancy rates had been advancing (from 7.6% in 1978 to 12.6% in 1998). This meant that continued oversupply contributed further to the erosion of property values in the 1990s (Ronald and Hirayama 2006).

Housing sector strategies were readdressed after 1997 when Japan's long recession was exacerbated by the Asian financial crisis. Housing policy was revised in order to address the necessity of economic restructuring and the demands of demographic change. Deregulation and marketisation became the focus of reforms with housing as a specific target. Most significantly, the GHLC was abolished in 2007 and replaced by the Housing Finance Services Agency (HFSA), leaving the private banking sector to fill the large gap in the primary loans market. In combination with the restructuring of government provision has been an equally fundamental restructuring in company society, undermining another pillar of the welfare mix. The government, under pressure from the company sector, began to support labour market casualisation after 1999. Many corporations subsequently abandoned the conventional system of lifelong employment and seniority pay, and many have also discontinued employee-housing and housing loan practices.

6.4.3 The Emerging Role of Housing in Social Insurance

Since 2001, government approaches to public policy and social insurance responsibilities have been remoulded, reflecting new socio-economic and demographic realities. On the one hand, the government abandoned the rhetoric of economic nationalism for a more neoliberal discourse focused on the deregulation and marketisation of government institutions and practices. On the other, following more than a decade of recession and marked demographic ageing, there has been a growing concern with the inadequacies of traditional welfare arrangements and mechanisms. 'Production-first' policy rationales have had diminishing impact on the electorate, and social policy issues have gained momentum. The outcome is an unusual combination of disengagement strategies (in the housing and pension sectors) and social policy initiatives (in public welfare) which aim to restructure the welfare mix around both individual provision and social care insurance.

A critical factor forcing this combination is the declining viability of the pension system. It became apparent in the early 2000s that the equilibrium rate of pension contributions and withdrawals would soon exceed the maximum possible rate of 21.6%, meaning that the system would be exhausted by 2050. From 2002, pension restructuring has meant that contribution levels could be increased and benefits reduced. The government specified in the 2004 reform that contribution rates would be raised every year until they reached 18.3% in 2017 but not raised thereafter.

At the same time, government commitment to old-age care has been radically augmented by the introduction of the long-term care insurance (LTCI) in 2000. This social insurance scheme provides access to at-home care services as well as health and welfare facilities (such as nursing homes) for older people. It marks the recognition of state responsibility for care of older people, which has previously been imposed on families. LTCI is non-means tested, ignoring differences in both contributions and the level of asset wealth or available family care, although there remains a focus on the family to provide care and the use of insurance only where it is absolutely necessary. A notable feature of state-subsidised services is the provision for home improvements for care receivers with government support for the remodelling of private homes to suit the mobility and care needs of older people.

While the state has taken on more accountability in providing for the growing older population, it can be expected that the ability of retired people to maintain their living standards in the future, in the light of expected declines in pension income (and the focus on at-home care services), will continue to depend on access to owner-occupied housing and the accumulation of housing assets. Due to the long post-war drive to expand homeownership, the vast majority of older people have become owner-occupiers whose properties can be used to offset income reduction as well as in the negotiation of intergenerational contracts (of health care and inheritance) within families (Izuhara 2007).

6.5 South Korea

Following the experiences of other high-growth economies where homeownership advanced rapidly, Korean governments have essentially assumed that as the stock of modern housing increased, a growing majority of families would become homeowners. Emphasis too has been placed on the family as the main provider of welfare care with ownership of a family home considered, along with market price appreciation, a primary means of providing security in old age. The state, meanwhile, has ensured intensive production of new units for sale for more than four decades: between 1962 and 2007, 15 million new homes were built of which 70.6% were constructed between 1989 and 2007 (KNHC 2009). However, homeownership rates remain modest and were, until the 1990s, effectively in decline (Park 2007).

The organisation of the housing system and the nature of government intervention, which has focused on supply rather than demand subsidy, have largely constrained the effectiveness of asset-based welfare in Korea. Though a significant number of Korean families have accumulated substantial property wealth and house price inflation has been high, access to home purchase has been polarised and the majority of urban households has struggled to get a foothold on the owner-occupied housing ladder.

The numbers of urban rental households and housing market conditions have shaped particular responses from governments in recent decades. Although the state has consistently emphasised the imperative of growth, interventions have often reflected social democratic policy features, especially during the Kim Dae-Jung and Roh Moo-Hyun administrations (1998–2008). A universal social insurance-type pension system has been in place since the late 1980s, although there have been adjustments in replacement rates as economic growth has slowed and demographic ageing advanced (Bonoli and Shinkawa 2005). European-like social security schemes have also been developed such as the National Minimum Livelihood Security Act of 2000. Although public rental housing provision has intensified, this has served the interests of the private construction sector and the needs of low-income working households in a highly stressed housing market.

6.5.1 The Housing System

Public sector housing development was initiated in the 1960s along with intensified state-guided industrialisation shaped by 5-year economic development plans. The Korean National Housing Corporation (KNHC) was established in 1962 as a self-financing public enterprise to construct new housing. Most public housing was for sale, however, and not targeted at low-income families. The Korean Housing Bank (KHB), which raised funds and lent money for home purchase through a contractual savings scheme, was also established in 1967.

State-directed housing provision took off in the 1970s along with large-scale development programmes involving private conglomerates in the mass construction of apartment blocks that became the mainstay of Korean urban housing. House prices began to escalate in the late 1970s as increasing numbers of better-off households speculated in new-build housing. The government responded with anti-speculation measures but, as market conditions cooled down, began to apply market stimuli. This pattern of speculative investment followed by a see-saw of reactionary market controls became characteristic of the Korean housing context.

Housing subsidies have historically focused on supply with provision for lower-income home buyers usually mediated through developers. Meanwhile, volatile markets have been the norm even though borrowing conditions have been restrictive. This combination has largely excluded many potential low- to medium-income buyers and constrained urban homeownership rates (Ronald and Jin 2010). Korean lending has been particularly cautious, and loan-to-value ratios of 30–40% and non-amortising 5–10-year loans are typical. Since the 1980s, the government has attempted to assist poorer home buyers directly through the National Housing Fund, although funds have been limited. Until the mid-1990s, the KHB still served as the primary financer in the mainstream mortgage market, accounting for around 75% of new housing loans. Since then, the KHB has been privatised, and there has been significant housing finance reform. In 2004, in order to facilitate greater flows of funds to middle-income households, the Korean Housing Finance Corporation (KHFC) was established. This has attempted to develop Korean mortgage practices by providing more flexible mortgage products.

Despite improved lending flows, the ownership of urban housing remains polarised, and in 2005, it was estimated that around 8% of households were multiple property owners and accounted for 38% of total housing stock (MGHA 2005). A contemporary landlord class has formed as a significant proportion of new housing has effectively become private rented housing. This sector is dominated by *chonsei* lettings in which tenants pay deposits of around 50–70% of the value of the property for, normally, a 2-year tenancy. This money is invested and later returned by the landlord with the interest made taken in lieu of rent. This system has been perceived to promote investment in housing and a means of forced savings that enables tenants eventually to buy a home.

The Roh government (2003–2008) attempted to address the inequalities of the housing system by squeezing multiple property owners through taxation. It also initiated a mass public rental housing programme that aimed to extend the sector from 1.5% to 17% of housing by 2017. Since 2008, the new, more socially conservative and economically liberal regime, under Lee Myun-Bak, has expanded the public housing program but included a large proportion of subsidised houses for sale.

6.5.2 Housing and Welfare

Despite the limited progress in expanding urban homeownership, this is still considered a key element of the welfare system. It has been argued that limited public service provision for older people has been based on the assumption that intergenerational family cohabitation provides the main pillar of welfare care (Park and Lee 2007). Recent policy reforms have strongly reflected this. In 1998, an ageing preferential deduction was established in income tax law, meaning that adult children living with their aged parents receive a tax deduction. An adult child that buys a home in which to cohabit with their elderly parent(s) also receives a transfer tax exemption, is given preferential access to special housing funds and, furthermore, can take higher loans for purchase or remodelling. More recently, and mirroring the Singaporean Lease Buyback Scheme, the Korean government has enabled the provision of reverse mortgage products that enable access to housing capital. Introduced with a government guarantee, this has been an explicit recognition of the potential of housing equity to meet pension needs (Doling and Ronald 2011).

Homeownership and intergenerational cohabitation thus form a key support for later life. Indeed, while urban homeownership rates are generally low, national rates are 76.3% among people of retirement age. However, this disguises significant inequalities and inadequacies of housing as a welfare and pension pillar. Firstly, older people are underrepresented in the modern urban housing stock. Secondly, homeownership among older people with low incomes is concentrated in poor-quality housing in rural and semi-rural areas. According to the Ministry of Construction and Transportation, while 16% of all houses fell below minimum housing standards in 2005, substandard housing accommodates 82.3% of low-income households (see Ronald and Jin 2010). Thus, not only is housing for older people often of poorer quality, it also functions poorly as a container of asset wealth as house price inflation is concentrated in the newer urban apartment stock.

South Korea provides some interesting contrasts to the Japanese and Singaporean cases. Although homeownership is considered a core of welfare security and while the state has pumped massive resources into the housing sector, the housing system and distribution of housing assets is largely unbalanced and inequitable. This may well have shaped public welfare demands and political pressures. It may also account for the more intensified provision of non-housing-based welfare goods and services in recent years and the ostensible extension of citizenship rights in terms of access to universal public benefits.

6.6   East Asian Insights

Although Japan, Singapore and Korea demonstrate very different approaches to homeownership provision and very different integrations of housing as a welfare pillar in the pension mix, together they illustrate the significance of the accumulation of individual or family housing wealth in meeting living costs in old age. They also provide insights into how asset-based welfare systems interact with changing social, economic and demographic conditions as well as what problems they generate for governments. Here, we assess these approaches against the adequacy and sustainability criteria.

6.6.1   Adequacy

Housing wealth is important in the pension mix in East Asian. It provides both a physical basis of the family model of welfare as well as a direct contributor to older people, for example, as a means of rent free living, an income in kind. Overall, the East Asian experience is that homeownership provides a supplementary tier of pension provision that may go a long way to maintaining the living standards retired individuals enjoyed during their working lives. Indeed, the size and quality of the owner-occupied dwelling may well be beyond what would be affordable in the market based on a pension income. However, homeownership is inadequate as the basis of retirement income and how effective it is as a supplement is dependent on the adequacy of first- and second-tier pension bases. Across East Asia, despite the emphasis on housing investment, other sources of income in old age can go much further to enhancing living conditions or preventing poverty.

Whereas the overall homeownership rate varies across the three countries examined, in all, the rate among older people is high. One consequence is that the proportion of the older population excluded from the benefits derived from homeownership is relatively small. But, in a number of other ways, the benefits are highly variable. In so far as this impacts on the adequacy of the income which older people enjoy, it points to some of the limitation of building homeownership as a pension pillar.

One dimension of the variability lies in the nature of housing stock itself. In Japan, for example, the longevity of built units is quite short, typically less than 30–40 years (see Ronald and Hirayama 2006). The maintenance cost of homes is thus also a factor in the performance of owner-occupation as a pension supplement. Similarly, in Korea, urban middle-class homeowners in newer properties enjoy the greatest benefits, while owners of cheap, older housing, especially in rural areas where wooden housing still dominates, may even be disadvantaged by the costs of home-ownership. In Singapore too, the deterioration of HDB apartment blocks has required the government to continue to subsidise renovation in order to maintain market values for public homeowners. Allied to this heterogeneity of the housing stock is also a heterogeneity in house value. In general, in all three countries, older people who own the most expensive homes have acquired them because they earned most in the labour market and will also tend to have the highest pensions and most other investments. Housing income in kind and in cash is thus heavily concentrated, tending to reduce the potential ability to compensate for low pensions or to reduce the risk of poverty, that is, to redistribute across income classes.

Finally, although housing asset wealth can, in principle, be realised as a cash value when required, it is vulnerable to market fluctuations. The vast majority of retirement savings of post-war generations have been redirected into housing properties, which may decline or go up in value at particular moments in the economic cycle. The living standards of retired people may thus be dependent on their equity stake in their home as well as its value at a particular point in the life course of the family. Property markets, timing of entry and movement up the housing ladder all play, therefore, a role in distorting the performance of housing assets as welfare supplements.

6.6.2   Sustainability

Historic market contingencies also create cohort differences between those who purchase homes at different times. It appears that the current assets built up by older post-war generations reflect historic housing market cycles in which housing became decreasingly affordable. For Groves et al. (2007), an unequal pattern is forming between different generations, with the gains made by older home purchasers increasingly difficult to reproduce. While the older generations who bought under favourable conditions (when prices were lowest and subsidies more advantageous) have experienced enormous increases, younger generations will, potentially, make much lower marginal gains and be more disadvantaged by market downturns.

In Japan, for example, substantial differences in housing equity and market position have been identified between baby boomer and post-baby boom (baby buster) cohorts (Hirayama and Ronald 2008). Exaggerated demand among baby boomers approaching early middle age drove the 1980s house price boom, making housing more expensive for the smaller cohorts that followed. The subsequent drop in demand undermined property values and had a particularly disproportionate effect on the capital gains of baby busters who bought at the price peak. These inequalities were enhanced by the economic recession that followed with negative equity being most prominent among households who bought their homes after the baby boomer wave.

Younger generations in Japan have been either more reluctant or less equipped to enter a more risky housing market, and homeownership rates have declined substantially among younger households. As average house price to income ratios remain high, while employment security has diminished, many younger households have found housing purchase increasingly unaffordable. Between 1978 and 1998, homeownership among those aged 25–29 fell from one in four to one in eight. For those aged 30–34, the drop was from one in two to one in five. Moreover, the Japanese housing sector is strongly polarised in terms of the size and quality of the stock. Consequently, a generational fragmentation is emerging between older owner-occupiers in spacious homes who rely for their pension incomes and welfare care services on the taxes and premiums of younger generations and young tenant households in small apartments unable to build up housing wealth through the owner-occupied housing ladder.

Over-investment in housing has become another issue. Homeownership subsidies and state facilities that speed up transition into homeownership (i.e. CPF, HDB, GHLC) have directed massive flows of savings into housing property investment. This has reduced investment in other kinds of assets and intensified pressure on urban housing markets and property prices. As socio-economic conditions changed, housing bubbles became increasingly vulnerable to volatile fluctuations, disrupting the welfare function of housing assets. This has become particularly evident in recent years following consecutive economic crises. While governments remain committed to supporting the asset wealth of existing homeowners, there have been necessary adjustments in constraints placed on housing investment vis-à-vis other savings vehicles (e.g. Singapore) and building up greater public welfare services for the elderly (e.g. Japan).

East Asian countries also illustrate the reliance of asset-based welfare systems on essentially unsustainable property value increases (Chua 2003). Governments have effectively become preoccupied with supporting of housing markets as declines undermine the national wealth built up in housing property as well as the ability of households to draw on housing assets in retirement. In Singapore, sustaining demand for owner-occupied housing units has become progressively problematic. A constant inflow of buyers is necessary to support property price increases. The government has repeatedly relaxed qualification criteria and extended subsidies, but there are now few categories of citizens remaining to extend access to HDB housing in order to invigorate demand. In Korea, meanwhile, governments have become preoccupied with stabilising the housing prices via various cooling and heating measures, leaving little room for the market to function.

Finally, changing inheritance patterns are also challenging the sustainability of the homeownership welfare pillar. With the decline in family size and the expansion of the number of older people, the chances of inheriting housing property have markedly increased. Nonetheless, the housing asset situations of younger generations are increasingly being determined by family conditions: whether or not their parents own a house, whether they can inherit it or not and whether or not they can obtain financial support from parents when purchasing a new home (Hirayama and Ronald 2008). In Singapore, Chua (2003) has explicitly identified that the increasing inheritance of homes, which will advance as cohorts of HDB homeowners age, will undermine the flow of house-purchasers, diminishing demand and generating a potential surplus.

6.7 Conclusions

The relevance of East Asian experience for considering the potential role of housing to meet the needs of older people in a European context lies in the former's intensity and longevity of homeownership policy pathways. In all three countries considered here, policies have promoted homeownership as a key component of their approaches to welfare providing one of the basic pillars supporting the well-being of individuals at all stages in the life cycle, as well as having a particular significance for meeting the needs of older people. At the same time, in terms of the embeddedness of the housing asset welfare pillar, they all constitute relatively mature systems. In Europe, by contrast, the role of housing assets has only relatively recently entered into policy debates. In this context, therefore, East Asia can be looked upon as providing Europe with an opportunity to learn lessons from those who have gone before.

Yet, there are marked differences between Europe and East Asia, not least of which is the under- or non-development of welfare states. The delayed progress in democratisation, citizenship rights and gender equality in East Asia need also to be taken into account. There would be dangers therefore in seeing the case studies as offering an opportunity for the direct transfer of policy. Rather, there are for the West lessons that have the potential to inform and enrich its policy debates.

The first of these lessons is that homeownership can play a pivotal role in protecting the well-being of older people. In the East Asian countries considered here the family, various forms of pension and savings, together with homeownership, have combined to create a mutually supportive system. The home may provide shelter and a physical focus for the family, but it also offers the possibility, if owned outright, for older people to live rent free, receiving an income in kind and thereby getting by on a smaller cash pension. In all three countries, governments have sought to reinforce this by providing opportunities and incentives to make it possible for different generations to live together or at least in close geographical proximity. More recently, there have also been policy developments establishing forms of reverse mortgage that enable access to housing assets, thus creating an income in cash. These countries, therefore, provide a clear example of an explicit targeting of housing equity by governments concerned about the viability of other forms of pension provision.

Notwithstanding these positive outcomes, a second message from East Asia is that homeownership may facilitate distribution across the life cycle and even across generations, but it is not generally a vehicle for distribution across income classes. In all three countries, the tendency in recent decades is for housing markets and property ownership to reinforce social inequalities rather than alleviate them. In general, those with least housing assets will have least non-housing income and wealth.

There are also lessons for Europe in regard to the sustainability of asset-based welfare systems. It has become clear that pre-1997 systems were built on the assumption that house price increases could outpace inflation in perpetuity. The reality of house price deflation results in part in additional pressure being placed on government to protect housing markets; house prices become an intensely political issue. In other words, governments cannot simply deflect responsibility for the well-being of older citizens by promoting homeownership, if that particular form of investment does not perform in a way that assures well-being.

It is also evident, however, that post-war generations of homeowners have been advantaged by their tenure status and that pro-homeownership policies, for a few decades at least, somewhat offset the underdevelopment of welfare programmes. Recent East Asian policy reforms, following the 1997 financial crisis, arguably reflect a new age in the development of homeownership and asset-based welfare characterised by attempts to offset over dependency on housing markets, on the one hand, and enhance the function and capacity of housing properties as welfare resources, on the other. Important, then, is the recognition that homeownership assets are not a complete substitute for forms of social provision. Homeownership is not sufficient on its own and, insofar as it works adequately to support people, does so in conjunction with, and not as a substitute for, other welfare pillars.

Chapter 7
Conclusions

7.1 Introduction

Housing equity is a pension; that is a fact: a large proportion of older Europeans own their own homes, mostly without outstanding housing loans, and they receive from that an income in kind in the form of rent-free living. Moreover, it has been received whether or not the acquisition of homeownership in the first place was motivated solely by consumption objectives and whether or not its investment potential has been explicitly recognised. While as income in kind it has enhanced living standards, homeownership has potentially offered a pension in another form also, as income in cash. This is not automatic, requiring the deliberate realisation of some or all of the equity, perhaps by moving to a cheaper house or taking out a reverse mortgage.

It is this potential for homeownership to offer not just a place for living but also a means of living that has formed the objective of the research underlying this book. Whereas there has been consideration of the interests of governments, the main concern has been with understanding how households have viewed this potential: how have they in the past used and sought to use housing equity to contribute to their income needs in old age, how are attitudes to its use developing and what would be some of the consequences of systematically using housing equity as a pension?

The aim of this final chapter is to summarise and reformulate the findings of our research into these questions. It starts with a summary of the general findings as they apply to Europe as a whole, focussing on the extent to which and how homeownership is used by households as a pension and to what extent this is fed or discouraged by a range of policies. The second section summarises the findings organised by the regime groupings in a series of five pen pictures. The intention here is to emphasise the variation across the member states in the experience and potential of housing wealth. The final part of the chapter discusses what the research findings might mean for policy in Europe.

J. Doling and M. Elsinga, *Demographic Change and Housing Wealth: Homeowners,*
Pensions and Asset-based Welfare in Europe, DOI 10.1007/978-94-007-4384-7_7,
© Springer Science+Business Media Dordrecht 2013

7.2 Accumulating and Using Housing Assets: A Summary of the Evidence

The research reported in the earlier chapters has been based on a range of methods and sources of data, which have, in turn, provided a range of perspectives. The limitations of each, however, have restricted the accuracy and completeness of their portrayal of the developments and outcomes across Europe. Nevertheless, in combination, they provide a set of pictures that offer some understanding of three key stages: the growth of homeownership across the member states, the role of housing assets in the wealth portfolios of European households and the use of those assets in later life. In turn, these understandings are broadly consistent with the notion of the mixed economy of welfare describing the formal institutions which, together with informal institutions, influence the choices made by European households.

7.2.1 The Growth of Homeownership

A necessary first step to accumulating wealth in the form of homeownership is obviously the acquisition of legal title, becoming a homeowner. Examination of trends over recent decades shows that increasing numbers of European households have been doing so. With some two-thirds of households now owning homes Europe as a whole can be described as a home owning society. At the same time, across member states, the popularity of homeownership varies considerably.

On the basis of both existing literature and new empirical evidence, it has been possible to identify what might underlie these trends. At a macro-level, there appears to be a link between the nature and extent of welfare provision and homeownership rates, reinforcing a notion that they are substitutable forms of horizontal distribution over the life cycle. Further, national housing policies appear to have made a difference and help to explain why homeownership rates differ so much over Europe. However, it is possible that it is not so much homeownership policies that make a difference but more the policies towards renting: those countries that have offered the opportunity to live in good quality rental housing with strong protection including regulated rents – which have also tended to be those countries with the most generous welfare systems – have experienced less growth in their homeownership sectors.

The interviews provide insights into how formal and informal institutions interlink. Buying a house at a young age is not the usual thing to do in Germany, for example, an outcome that can be attributed to the strict criteria of lenders that in practice exclude younger households, but also by the risk and debt averseness of German households who have the habit of saving before buying. This is made possible by the existence of a large private rental market of good quality and security of tenure. More generally, it seems that in countries with particularly large homeownership rates, this form of tenure is so embedded into the way of life that it has become the natural and normal option. Policies or formal institutions may seem to move in the same direction, therefore, but informal institutions appear to make a difference.

From the micro-statistical level, income appears as an important factor, with higher-income groups being more likely than low-income groups in all countries to be owners. Age also appears important, and although it is difficult to disentangle age and cohort effects, the general pattern is of homeownership rates to increase during working years and to decrease in older age groups: an inverted U-shape which is consistent with the LCM.

7.2.2   The Accumulation of Housing Wealth

While trends in rates of homeownership can be relatively accurately identified, information about the amounts and distributions of housing wealth is not so clear. Nevertheless, from the limited, harmonised data, it does appear that the total amount of housing wealth is considerable, perhaps equivalent to one and a half times the combined GDP of all the member states, while it also constitutes the largest single form of wealth for the average European household, particularly so for older Europeans. It also appears more evenly distributed across populations than some other forms of wealth such as shares. In a numerical sense, therefore, housing wealth appears to be significant.

But, understanding why the absolute amount of average housing wealth, as well as its size relative to other forms of household wealth, varies across member states is less certain. There appear to be a number of influences although their relative importance cannot be specified. Firstly, based on a very small number of countries – four European and one non-European – it is possible that higher levels of government welfare spending are associated with a lower level of non-housing assets in the household portfolio. The limited data basis should prevent any strong conclusions, but these findings suggest the possibility of a trade-off between welfare expenditure and homeownership may need to be qualified: the response to a lack of generosity in state welfare spending may not necessarily be the acquisition of more housing wealth but of more wealth in forms that may be more easily realisable.

Secondly, the interviews revealed that in some countries, tax policies play an important role in household strategies: for example, if homeowners can deduct mortgage interest, they benefit from having a housing debt and therefore build less housing equity. Moreover, formal and informal institutions interact as is illustrated by the Dutch case. The generous mortgage tax relief turned a mortgage debt from 'preferably as small as possible' in the Calvinistic Netherlands of the 1950s into a regular part of a smart household portfolio in the 1990s.

Thirdly, our macro-analysis revealed a link between the returns on different types of asset and their relative size in the wealth portfolio: simply, the higher the return, the higher the proportion of the portfolio. Whereas this holds for stocks and shares, it is less pronounced for housing as a financial asset, one reason for which could be the modifying effect on household behaviour of the non-financial importance of housing.

Fourthly, and complicating this interpretation, the interviews revealed that both formal and informal institutions play an important role in household strategies. Some institutions, such as the value that buying 'a house is a good investment', are similar in most countries. Although buying a house may in the first instance be a matter of consumption, over time households often come to realise that housing is also an investment. Buying a dwelling is a means of self-discipline that prevents households from spending too much and a semi-conscious pension strategy feeding the informal institution that 'housing is a good investment'.

Finally, it is clear that the expansion of homeownership sectors has been accompanied by expansion of housing finance sectors. While for most European households access to homeownership is made possible with a loan or financial support from the family, there appear to be large cross-country differences in attitudes to debt. In corporatist (Germany and Belgium) and Mediterranean (Portugal) countries, for example, households often emphasised that it is very important for them to pay off debts as soon as possible.

7.2.3 The Use of Housing Wealth in Old Age

In principle, there are a number of ways in which households could dissave. Whereas income in kind is a benefit for all homeowners, realising housing equity in order to create income in cash requires deliberate action, involving either moving to a cheaper house or taking a financial product such as a reverse mortgage. As with the accumulation of housing assets, however, the study of the decumulation, or dissaving, of housing assets is similarly beset by inadequacies in publishable data. Nevertheless, some general patterns of behaviour are identifiable. In broad terms, European households appear to reduce their total wealth progressively through their retirement years, in effect enhancing their pensions and in that way acting consistently with the LCM. At the same time, the dissaving of housing assets does not appear routinely pursued.

The widespread reluctance to realise housing equity begs the question: why do Europeans realise and spend non-housing assets while holding on to their housing assets? There appear to be a number of reasons. One reason may lie in the fact that housing is both a consumption and an investment good. A consequence is that one way in which older households benefit financially from their tenure status is by having low housing expenses, in effect receiving an income in kind which allows them to get by on a smaller pension. Another consequence, however, is that any realisation of a lump sum or income in cash may impact on the flow of housing services. While there are a number of ways of avoiding this – from becoming a renter to taking a reverse mortgage product – the most commonly adopted strategy appears not to dissave housing assets at all but, as far as housing is concerned, to maintain the status quo.

This general reluctance to spend housing assets may arise because, notwithstanding widely held views about the inadequacy of pension systems and their individual pension entitlements, most older Europeans actually have sufficient income not to have to 'spend' their housing. However, there appear to be other reasons. For many, housing is viewed as a substitute not only for the inadequacies of government pension provision but for government provision in other areas, especially social and health-care needs. For many, housing assets seemed to be retained as a precaution or safety net against future events.

But for many, the home has meanings quite separate from its ability to finance welfare needs. The home as a bequest, sometimes physical sometimes financial, to be left to children is clearly an important consideration. This is perhaps no more so than in countries in which the home is a central orientation of the extended family, which itself is a vehicle for horizontal distribution. Such motivations appear to inform not only the general tendency to hold onto housing assets but also the particular lack of a buoyant demand for reverse mortgage products.

Releasing equity by mortgage equity release schemes is far from self-evident, then. The precautionary and the bequest motives are accompanied by informal institutions: 'saving is good and debts are bad,' and the notion of a debt-free, ownership ideal is strong playing an important role in many countries, particularly for older generations. Moreover, 'you should leave something for your children' is the norm among people with children, even if children are well off. Finally, deep 'distrust against financial institutions' seems everywhere significant.

There also appears to be a variation, common across countries, in generational attitudes. The pre-baby boom generations with memories of austerity and hardship are often very cautious, reluctant to spend their assets on consumption and eager to pass on an inheritance to their children. Commonly, baby boomers and later generations appear much more willing, sometimes eager, to continue, if not increase, the level of consumption they had enjoyed while working, and if this could be achieved by using the equity in their home that was acceptable. Younger age groups often appear even more open to the necessity to have to find their own solutions to their income needs in older age and to use housing assets to do so. Of course this might also be an age rather than a cohort difference, but it indicates the possibility that past attitudes and behaviour may not simply roll into the future.

7.3   Differences Between Welfare Regimes

Whereas the previous section provides a summary of the overall pattern of behaviour across Europe, our evidence has also indicated distinctive patterns in each of the five welfare regimes. Although these also mask some within-group, that is country to country, variation, in the main this seems, at least for most of the groups, less significant than the between-group variation. These are important in helping to generalise the understanding of housing assets in old age across the full range of the member states. Table 7.1 summarises key characteristics that map out different contexts in which the use of housing equity in old age has developed.

Table 7.1 Characteristics of regime groupings

	Corporatist	Social democratic	Mediterranean	Liberal	Eastern
Homeownership rate	Low	Low	High	Medium	Very high
Mortgage lending as % GDP	Low	High	Mixed	High	Low
Welfare spending as % GDP	High	High	Low	Low	Low
Concern about adequacy of pensions	Medium	Low	High	High	High
Homeownership wealth as % GDP	Low	Low	Mixed	Medium	High
Mobility among older owners	Low	High	Low	High	Low

7.3.1   Eastern: Housing as Ultimate Precautionary Fund I

Generally in these countries, homeownership has grown to very high rates largely as a result of the economic and political transformations taking place over the last two decades rather than the development of mortgage markets. Welfare spending generally is low, and large proportions of their populations do not consider their pension systems to be safe. In these circumstances, housing equity has become the ultimate precautionary fund. The family plays an important role with adult children often providing support to their parents with the understanding that the family home will become theirs: in that way, housing equity is part of the family strategy. Indeed, equity release arrangements are possible through non-profit organisations (Slovenia) or local government (Hungary), providing households with an alternative to support through the family, but, despite their strong promotion in some countries, markets remain small.

7.3.2   Mediterranean: Housing as Ultimate Precautionary Fund II

High rates of homeownership have a long tradition in these countries based on its role as a focus of the family. In recent decades, mortgage markets have expanded, particularly in Spain, and rates of outright ownership by young people tend to be low. In other ways, the Mediterranean countries have much in common with the Eastern regime countries. Although pension systems are very generous in some Mediterranean countries (e.g. Italy), in many, they are not considered safe. Housing equity is considered important, becoming the ultimate precautionary fund. Generally, older people do not move to cheaper homes or into rental housing, reflecting the central position of the home as something not to be treated as a financial commodity but as part of a wider family strategy.

7.3.3   Liberal: Shift from State to Housing Equity

Homeownership rates have increased in recent decades fuelled to a large extent by the expansion of mortgage markets. Pensions and the welfare system in general have been under pressure, there is large distrust of private pension funds, and housing equity is considered an important asset by households. There is a widespread view of housing as a financial investment; many households have built housing equity, and many also consider realising the equity by moving or by using equity release products. But, there is also concern about state protection in general. Although the equity release market in the UK is large in comparison with the other member states, even so such products are used by only a small proportion of older people.

7.3.4   Corporatist: Shift from State to Family

Though mandatory pensions are relatively generous in the corporatist countries, many households are worried about income in old age. In both Germany and Belgium, housing income in kind is considered an important part of income, while equity release products are not considered attractive options. On the contrary, households prefer to pay off the mortgage as soon as possible in order not to be indebted. The family and the bequest motive play an important role here. In some countries, especially Germany, not only inheriting owner-occupied dwelling but also ownership of rental dwellings is often part of the family strategy. Therefore, it seems that the family rather than the financial markets has been the more important in redistributing housing income over the life cycle.

The Netherlands, although otherwise classed as a corporatist country, shows many similarities to Sweden and Denmark. In the Netherlands, households consider the public as well as the occupational pension system as relatively reliable. Fiscal policy encourages people not to build housing equity or at least not to pay off their mortgages so that housing equity constitutes a relatively modest part of the total asset portfolio.

7.3.5   Social Democratic: Modest Role for Housing Equity

Both Sweden and Denmark, the archetypal social democratic countries, have large rental markets and high levels of mortgage debt. With relatively generous welfare systems offering considerable protection for their citizens, housing equity does not play an important role: there are fewer homeowners, and homeowners build less housing equity. As in the Netherlands, many people extend their housing loans into their retirement years. In some respects, Finland differs, having a larger homeownership sector and less trust in its pension system.

7.4 Policy Matters

Finally, then, what do our findings indicate that might inform debates about the role of homeownership as a solution to the so-called pension crisis? One response to this lies in the evidence from our findings that enables evaluation of the effectiveness of housing equity, were it to be fully and routinely used to provide older people with an income. A second response considers a number of wider issues that would need to be addressed by policy makers if their objective was to achieve a position where households would be willing and able to use housing assets in this way. This is not a proposition that such an objective is necessarily desirable or recommended, but setting it up as if it were provides a context for identifying what some of the implications might be.

7.4.1 How Would Housing Perform as a Pension?

While there may be a number of ways in which to evaluate how well pension systems work, the approach widely adopted in European policy analyses has been to use the adequacy and sustainability criteria.

7.4.1.1 Maintaining Former Living Standards

There can be little doubt that homeownership already, that is, even without further developments, makes an important contribution to maintaining former living standards. People who are outright owners of their homes are able to live rent free. In contrast to those who pay rent, they receive a net income in kind, the amount of which varies according to the value of the home. The average, in comparison with cash income from pension and savings, varies from member state to member state, but mostly appears to add between a quarter and a half to their other sources of income. If all older homeowners also drew directly on the housing equity, say through a reverse mortgage product, this would, very approximately, contribute the same sort of addition. So, in total from housing income in kind and housing income in cash, older people would get between an extra 50% and 100% increase in their overall income.

While this indicates that housing wealth could on average significantly enhance other household income and thus increase their standard of living, there is an issue of fundamental importance in assessing the actual effects: would the housing income in cash be an *addition* or *complement* to existing sources of income or a *substitute* for that part formerly derived from the state in the form of a pension? As a *complement*, housing wealth offers the potential to older people to considerably boost their cash income and thereby enhance their living standards. Quite simply, by cashing in some of their assets, they can consume more. As a *substitute*, however, governments may use the potential of housing wealth as a rationale for reducing the value of the state pension (or other state spending on older people). The consequences could include people being 'forced' to use their housing wealth even if they would prefer to leave a bequest to their children, perhaps leading to some erosion of solidarity both across generations and across society. Furthermore, in this scenario, those who rent their homes might be doubly disadvantaged in having both a reduced state pension and no housing assets to draw upon.

7.4.1.2   Preventing Poverty

Even if housing assets are treated as an addition to existing pension arrangements, another concern is whether the addition would lift older people out of poverty. Of course, by definition, housing wealth cannot help those people who do not have housing wealth to begin with – broadly, the 25% of older Europeans who rent their homes. In fact, in most member states, renters will tend to be poorer than owners and therefore most at risk of poverty anyway. To that extent, the use of housing equity might do little to ameliorate those who have the lowest incomes.

But even among those who do own their own homes, those with least pension and savings income tend to own the cheapest homes and therefore benefit from the least income in kind and potential income in cash. Another way of putting this is that housing would make a large contribution to reducing the risk of poverty where people who had low incomes held a lot of housing wealth, the so-called income-poor, asset-rich. Our evidence suggests that whereas there are some income-poor, asset-rich older Europeans, the general picture is that older people who have a lot of housing equity also tend to have had higher incomes when working, to have accumulated large non-housing assets and to have higher pensions. The general picture is of a positive correlation between income, housing assets and non-housing assets.

Existing pension systems across the European member states commonly have a significant first-tier element which effectively provides protection for those with least pension and other resources. Whether or not that protection was retained, the use of housing assets would do little to reinforce that protection. The evidence from the East Asian experiences is that homeownership may facilitate distribution across the life cycle, but it is not generally a vehicle for distribution across income classes. Indeed, the tendency in East Asia in recent decades has been for housing markets and property ownership to reinforce social inequalities rather than alleviate them.

7.4.1.3   Providing a Sustainable Pension

Whereas in the past the investment potential of homeownership may have made it an attractive option to many European citizens, and thus contributed to the growth of homeownership sectors, that growth was frequently enhanced by state policies such as subsidies. This raises the issue of whether the size of homeownership sectors will be maintained. The future of such policies will in part be contingent on the desire of member state governments further to promote homeownership. Such desire can be expected to be influenced by a number of the economic and fiscal considerations: concerns about maintaining the growth and stability conditions on which inclusion in the eurozone is dependent, on the consequences of the present economic and financial crisis and on the changing dependency ratios that are the basis of the pension crisis.

Although at the time of writing it is not clear where current developments are leading, it seems a justifiable assessment that any failure to sustain even the existing size of homeownership sectors along with the growth in house prices will result in a reduction in the potential of housing equity to meet income needs. There are, here also, relevant lessons from East Asia in regard to the sustainability of asset-based welfare systems. It has become clear that pre-1997 systems were built on the assumption, that has proved to be false, that house price increases can outpace inflation in perpetuity. Also evident have been the destabilisation effects of homeownership-dependent housing and welfare systems. Marked generational divides have illustrated that favourable conditions for market access and equity accumulation cannot simply be reproduced from one generation to the next. A similar development seems to be going on in Europe now, the younger generation may well not benefit from the housing market as the generation entering the market before the current financial crisis did.

It is also evident from the East Asian experience that post-war generations of homeowners have been advantaged by their tenure status and that pro-homeowner-ship policies, for a few decades at least, did somewhat offset the underdevelopment of welfare programmes. The problem for these cohorts has been the liquidity of housing assets when needed to serve individual welfare consumption needs. Recent East Asian policy reforms appear to reflect a new age in the development of home-ownership and asset-based welfare characterised by attempts to offset over dependency on volatile housing markets, on the one hand, and enhance the function and capacity of housing properties as welfare resources, on the other. While the former has required housing system diversification, the latter has called for innovations in types of public equity release schemes. This is an interesting development that is relevant for Europe; however, it is too early to draw a conclusions on the long-term sustainability of such public or public guaranteed equity release schemes.

7.4.2 Wider Policy Issues

Whatever the view taken of the extent to which housing assets measure up to the adequacy and sustainability criteria, and of the desirability of restructuring existing pension systems to incorporate housing as a central pillar, there are a number of additional implications for policy makers. Since pension and welfare systems are under pressure, the role of housing equity will undoubtedly become more important, and one option is for Europe to develop a good functioning equity release market. Nevertheless, achieving a situation in which housing assets are routinely used to augment income in old age would be expected to pose a number of challenges.

7.4.2.1   Persuading Households to Include Housing Equity
in Their Strategy for Old Age

Housing may be considered a pension by many, but why do households not release equity to the extent predicted by an LCM perspective or required by a reconfiguration of pension systems that rest on a central role for housing assets? There are a number of barriers. Firstly, housing has a special role for many households, especially where it is seen as part of the family project. As Reifner et al. (2009) argue, one of the first steps to the mass expansion of reverse mortgage markets is an acceptance that housing is not just a consumption good with emotional and cultural attributes but also a financial vehicle. This is a barrier especially in the Mediterranean countries. In many circumstances, this is tied also to the notion of the home as a bequest, part of the contract between family generations, so that housing equity even without being released plays a key role in pension and welfare provision for households.

Secondly, housing equity is seen as a safety net, as a last option in the case of urgent need. Households appear often to cherish this last option in particular when the future of welfare provision is insecure; care in old age is a great source of worry, and implicitly households see releasing housing equity as last resort. There may be a paradox here. Households in countries with generous welfare systems accumulate less housing equity in the first place but appear less concerned about being outright owners. This appears the case in Denmark and the Netherlands, for example. But, households in countries with less generous welfare systems accumulate more housing assets, more often seek to own them outright and are more reluctant to realise them. Thus, the release of housing equity appears conditional on, and not a general substitute for, state welfare spending. Another way of putting this is that government efforts to promote housing as a pension may founder if governments see that as a part of a general withdrawal from social provision.

Thirdly, where older people do want to use their housing equity, the strategy of doing so by selling one's home and seeking to move into rental housing, requires a supply of suitable rental housing. This is not only a matter of the size of rental sectors but also the regulatory regime which protects the interests of tenants, the controls of rents which may make them affordable and the physical characteristics of rental properties fitting the specific needs of older people.

7.4.2.2   Persuading Financial Institutions to Offer Sustainable
and Transparent Equity Release Products

As detailed in Table 4.2, realising housing equity to create an income in cash can be achieved in a number of ways. Insofar as it seems unlikely that large proportions of older people in all member states will want to remain in their home, rather than move down market or out of market, this would require the extension of reverse mortgage solutions. In many member states, this extension would be from very low, even practically nonexistent, starting points. The EU's Green Paper question of whether it would be useful, through its single market objectives, to extend the market in reverse mortgages might be a necessary first step, but there are other, perhaps more difficult, constraints.

For financial institutions in general, especially because of the limitations on their lending behaviour as a consequence of the current financial crisis, the reverse mortgage market is not necessarily deemed a priority. Reverse mortgages require the advancing of large amounts of money, which will be repaid at an undefined point in the future, during which period there will be changes in interest rates and the value of the house from which final payment will be made. In East Asian countries that have sought to extend access to reverse mortgages, these risks have been obviated by government intervention: in the case of Singapore by the government itself buying back years of the leasehold and in Korea by the government providing guarantees, thus protecting financial institutions against house price risk and thereby enabling them to offer financially more attractive terms to households. The lesson here of course is that establishing mass access to reverse mortgage opportunities has not occurred spontaneously, and would almost certainly require state support, so that it would not be fiscally costless and would imply substantial risk for governments.

7.4.2.3   Supporting Investment Decisions

There is an issue about how governments support older people to make wise investment decisions. In a world in which pension provision is organised through the state (and perhaps employers), expert advice can be utilised in order to get appropriate returns, and where underperformance, in terms of adequate provision, is met, the state's resources may obviate them. But, if the pension challenge becomes individualised including an expectation that people will use their housing wealth to fund their pensions, how is expert knowledge acquired? People often do not have much time or knowledge that enables them to compare all details of financial products. It would be important that there is not only adequate supervision of financial markets but also adequate information and advice on different options to support households in making their choices.

7.4.2.4   How About Renters and Rental Housing?

While the status of homeownership may provide a solution to some people's pension needs, it leaves a key problem for governments of how to deal with tenants, often the lower-income groups. But, it might mean establishing protection also for the owners of homes with market values so low that they would not yield an income in cash large enough to protect them from the risk of poverty. These needs would almost certainly entail a first-tier pension scheme which can provide some vertical redistribution. Specifically, there may be a moral hazard problem: how can governments both encourage households to take care of themselves while, at the same time, if they are to sustain a commitment to creating a decent and inclusive society, making support available in case of need.

Encouraging a housing sector for older people including care arrangements is another option. For homeowners, moving to a rental dwelling purpose-built for older people is an opportunity to turn housing equity into care. This might be particularly applicable in circumstances where a barrier to realising housing capital is the belief that housing equity must be kept as a precaution against the cost of care in old age. However, this option requires investment in the rental sector.

7.4.2.5   Is Housing Equity an Alternative to Taxation?

For governments seeking to grapple with fiscal challenges, resulting not only from demographic ageing but also the current financial crisis, the wealth embedded in homeownership might well seem to offer a possible solution: private saving leading to private wealth as an alternative to taxation and public expenditure. There are a number of reasons why these might not create the desired fiscal payoff. To those identified above – maintaining welfare safety nets and supporting reverse mortgages – can be added another. Homeownership may be viewed as a form of forced saving. The greater the number of people who meet their housing needs by becoming homeowners, the more people are also investing in a capital asset. The evidence reported in this book indicates however that people make trade-offs between different means of achieving horizontal distribution across their life cycles. It is possible therefore that additional saving through housing markets will be offset by reduced saving in private, that is tier 3, pension schemes or other assets. Greater reliance on housing asset, therefore, would not necessarily be part of an overall shift towards private rather than public pension arrangements.

7.4.2.6   Responsibility for Housing Market Outcomes

Like all asset classes, housing is risky. Although as the evidence in Chap. 3 indicated house prices tend to be less volatile than share prices, they may nevertheless go up or down. The experience from East Asia is that in situations where the well-being of people is based, in part, on the value of their homes and the government is party to that arrangement, the government also bears a responsibility for ensuring that housing markets behave in ways that do not jeopardise house values. In addition to the fiscal costs of supporting the wider use of housing equity, then, are political risks tied to the protection of the electorate against market developments that would adversely affect them.

References

Albertini, M., Kohli, M., & Vogel, C. (2007). Intergenerational transfers of time and money in European families: Common patterns – Different regimes? *Journal of European Social Policy, 17*(4), 319–324.

Allen, J., Barlow, J., Leal, J., Maloutas, T., & Padovani, L. (2004). *Housing and welfare in southern Europe*. Oxford: Blackwell.

Altissimo, F., Georgiou, E., Sastre, T., Valderrama, M., Sterne, G., Stocker, M., Weth, M., Whelan, K., & Willman, A. (2005). *Wealth and asset effects on economic activity* (Occasional Paper Series No 29). Frankfurtam am Main: European Central Bank.

Andrews, D., Caldera Sánchez, A., & Johansson, A. (2011). *Housing markets and structural policies in OECD countries* (Economics Department Working paper No 836). Paris: OECD.

Apgar, W. C., & Xiao Di, Z. (2006). Housing wealth and retirement savings: Enhancing financial security for older Americans. In G. L. Clark, A. H. Munnell, & J. Michael Orszag (Eds.), *Oxford handbook of pensions and retirement income*. Oxford: Oxford University Press.

Atterhög, M. (2005). *Importance of government policies for home ownership rates* (Working Paper, No 54). Stockholm: Royal Institute of Technology, Section for Building and Real Estate Economics.

Attias-Donfut, C., Ogg, J., & Wolff, F. C. (2005). European patterns of intergenerational transfers. *European Journal of Ageing, 2*(3), 161–173.

Balchin, P.N. (1996). *Housing policy in Europe*. London: Routledge.

Boelhouwer, P., & Van der Heijden, H. (1992). *Housing systems in Europe: Part I: A comparative study of housing policy*. Delft: Delft University Press.

Bonoli, G., & Shinkawa, T. (2005). Population ageing and the logics of pension reform in Western Europe, East Asia and North America. In G. Bonoli & T. Shinkawa (Eds.), *Ageing and pension reform around the world: Evidence from eleven countries*. Cheltenham: Edwin Elgar.

Bonvalet, C., & Ogg, J. (2008). The housing situation and residential strategies of older people in France. *Ageing and Society, 28*(6), 263–285.

Boone, L., & Girouard, N. (2002). The stock market, the housing market and consumer behaviour. *OECD Economic Studies, 35*(2), 175–200.

Bridges, S., Disney, R., & Henley, A. (2004). Housing wealth and the accumulation of financial debt: Evidence from UK households. In A. Bertola, R. Disney, & C. Grant (Eds.), *The economics of consumer credit*. Cambridge: MIT Press.

Calza, A., Gartner, C., & Sousa, J. (2001). *Modelling the demand for loans to the private sector in the Euro Area* (Working Paper 14). Frankfurt am Main: European Central bank.

Campbell, J. (2006). Household finance. *Journal of Finance, 61*, 1553–1604.

Castles, F. G. (1998a). The really big trade-off: Home ownership and the welfare state in the new world and the old. *Acta Politica, 33*(1), 5–19.

Castles, F. G. (1998b). *Comparative public policy: Patterns of post-war transformation.* Cheltenham: Edward Elgar.

Castles, F. G. (2004). *The future of the welfare state.* Oxford: Oxford University Press.

Castles, F. G., & Ferrera, M. (1996). Home ownership and welfare: Is southern Europe different? *South European Society and Politics, 1*(2), 163–185.

Catte, P., Girouard, N., Price, R., & Andre, C. (2004). *Housing markets, wealth and the business cycle* (Economics Department Working Paper No 194). Paris: OECD.

Cecodhas. (2008). *Review of social housing in Europe.* Brussels: Cecodhas.

Chua, B. H. (2003). Maintaining housing values under the condition of universal homeownership. *Housing Studies, 18*(3), 765–780.

Chiuri, M., & Jappelli, T. (2006). *Do the elderly reduce housing equity? An international comparison* (Working Paper No 158). Salerno: Centre for Studies in Economics and Finance, University of Salerno.

Clark, W., Deurloo, M., & Dieleman, F. (1994). Tenure changes in the context of micro-level family and macro-level economic shifts. *Urban Studies, 31*, 137–154.

Clerc-Renaud, S., Pérez-Carrillo, E. F., Tiffe, A., & Reifner, U. (2010). *Equity release schemes in the European Union.* Norderstedt: Books on Demand.

Coleman, D. (2001). Population ageing: An unavoidable future. *Social Biology and Human Affairs, 66*, 1–11.

Conley, D., & Gifford, B. (2006). Home ownership, social insurance and the welfare state. *Sociological Forum, 21*(1), 55–82.

Creswell, J., & Plano Clark, V. (2007). *Designing and conducting mixed methods research.* London: Sage.

Deaton, A. (1992). *Understanding consumption.* Oxford: Clarendon Press.

Demography Report. (2008). *Meeting social needs in an ageing society* (Commission Staff Working Document).

Disney, R., & Johnson, P. (2001). *Pension systems and retirement incomes across OECD countries.* Cheltenham: Edward Elgar.

Disney, R., Johnson, P., & Stears, G. (1998). Asset wealth and asset decumulation among households in the retirement survey. *Fiscal Studies, 19*(2), 153–174.

Doling, J. (1997). *Comparative housing policy.* Basingstoke: Macmillan.

Doling, J., & Ford, J. (2003). *Globalisation and home ownership: Experiences in eight member states of the European Union.* Delft: Delft University Press.

Doling, J., & Ford, J. (2007). A union of home owners, editorial. *European Journal of Housing Policy, 7*(2), 113–117.

Doling, J., & Horsewood, N. (2003). Home ownership and early retirement: European experience in the 1990s. *Journal of Housing and the Built Environment, 18*, 289–308.

Doling, J., & Ronald, R. (2010). Home ownership and asset-based welfare. *Journal of Housing and the Built Environment.* Published online an January 10, 2010, http://www.springerlink.com/content/535173346479q172/fulltext.pdf

Doling, J., & Ronald, R. (2011). Meeting the income needs of older people in East Asia: Using housing equity. *Ageing and Society, 31*, 2011.

Dübel, A. (2008). German Subprime Meltdown? *Interview with Achim Dübel.* http://www.ritholtz.com/blog/2009/05/german-subprime-meltdown-interview-with-achim-dubel/

ECB. (2003). *Structural factors in the EU housing markets.* Frankfurt: European Central Bank.

ECB. (2009). *Housing finance in the EURO area* (ECB Occasional Paper No. 101).

Elsinga, M., & Hoekstra, J. (2005). Home ownership and housing satisfaction. *Journal of Housing and the Built Environment, 20*(4), 401–424.

Elsinga, M., & Mandic, S. (2010). Housing as a piece in the old age puzzle. *Teorija in praksa: Revija za druzbena vprasanja, 47*(5), 940–958.

Elsinga, M., de Decker, P., Toussaint, J., & Teller, N. (Eds.). (2007). *Beyond asset and insecurity: On (In)security of home ownership in Europe.* Amsterdam: IOS Press.

Elsinga, M., Haffner, M., & van der Heijden, H. (2008). Threats for the Dutch unitary model. *European Journal of Housing Policy, 8*(1), 21–37.

EMF. (2010). *Hypostat*. Brussels: European Mortgage Federation.

Ermisch, J., & Jenkins, S. (1999). Retirement and housing adjustment in later life: Evidence from the British household panel survey. *Labour Economics, 6*, 311–333.

Esping-Andersen, G. (1990). *The three worlds of welfare capitalism*. Cambridge: Polity Press.

EU. (2006, December 22). *Report of the Mortgage Funding Expert Group*. Brussels: European Commission Internal Market and Services DG.

EU Housing Ministers. (1999). *Final Communiqué*. 11th informal meeting of EU Housing Ministers, Kuopio, Finland.

European Commission. (2005a). *Green paper. Confronting demographic change: A new solidarity between the generations* (COM (2005) 94 final). Brussels: Commission of the European Communities.

European Commission. (2005b). *Public finances in EMU – 2005*. Luxembourg: Office for Official publications of the European Communities. http://www.europa.eu.int/comm/economy/publications/publicfinance_eu.htm

European Commission. (2006). *Adequate and sustainable pensions, synthesis report* (Directorate General for Employment, Social Affairs and Equal Opportunities). Luxembourg: Office for Official Publications of the European Commission.

European Commission. (2009). *Eurobarometer*. Brussels: European Commission.

European Commission. (2010a). *Communication from the commission: Europe 2020: A strategy for smart, sustainable and inclusive growth*. Brussels: European Commission.

European Commission. (2010b, July 7). *Green paper: Towards adequate, sustainable and safe European pension systems* (SEC(2010)830). Brussels: European Commission.

European Communities. (2004). *Facing the challenge: The Lisbon strategy for growth and employment*. Report of the High level Group chaired by Wim Kok. Luxembourg: Office for Official Publications of the European Communities.

Eurostat. (2008). *European Union Statistics on Income and Living Conditions*. Brussels: Eurostat.

Evers, A., & Laville, J.-L. (Eds.). (2004). *The third sector in Europe*. Cheltenham: Edward Elgar.

Fahey, T. (2003). Is there a trade-off between pensions and home ownership: An exploration of the Irish case. *Journal of European Social Policy, 13*(2), 159–173.

Feijten, P., & Mulder, C. (2002). The timing of household events and housing events: A longitudinal perspective. *Housing Studies, 17*(5), 773–792.

Feinstein, J., & McFadden, D. (1987). *The dynamics of housing demand by the elderly: Wealth, cash flow, and demographic effects* (NBER Working Papers 2471). Cambridge, MA: National Bureau of Economic Research, Inc.

Follain, J. R., & Dunsky, R. M. (1997). The demand for mortgage debt and the income tax. *Journal of Housing Research, 8*(2), 155–199.

Folster, S. (2001). Asset-based social insurance in Sweden. In S. Regan & W. Paxton (Eds.), *Asset based welfare: Interactional experiences* (pp. 74–85). London: IPPR.

Forrest, R., & Hirayama, Y. (2009). The uneven impact of neo-liberalism on housing opportunities. *International Journal of Urban and Regional Research, 33*(4), 998–1013.

Frericks, P. (2010). Capitalist welfare societies' trade-off between economic efficiency and social solidarity. *European Societies, 12*(5), 719–741.

Frick, J., & Grabka, M. (2002). *The personal distribution of income and imputed rent: A cross-national comparison for the UK, West Germany and the USA* (Discussion Paper 271). Berlin: DIW Berlin, German Institute for Economic Research.

Gibb, K. (2002). Trends and change in social housing finance and provision within the European Union. *Housing Studies, 17*(2), 325–336.

Glennerster, H. (2003). Paying for welfare. In P. Alcock, A. Erskine, & M. May (Eds.), *The student's companion to social policy* (2nd ed., pp. 253–259). Oxford: Blackwell.

Groves, R., Murie, A., & Watson, C. (2007). *Housing and the new welfare state: Examples from East Asia and Europe*. Aldershot: Ashgate.

Gruis, V. H., & Priemus, H. (2008). European competition policy and national housing policies; international implications of the Dutch case. *Housing Studies, 23*(3), 485–505.

Has the rise in debt made households more vulnerable? *OECD Economic Outlook*, No. 80, 135–158.

Hills, J. (1993). *The future of welfare: A guide to the debate.* York: Joseph Rowntree Foundation.

Hirayama, Y. (2003). Housing policy and social inequality in Japan. In M. Izuhara (Ed.), *Comparing social policies: Exploring new perspectives in Britain and Japan.* Bristol: Polity Press.

Hirayama, Y. (2010). The role of home ownership in Japan's aged society. *Journal of Housing and the Built Environment, 25*(2), 213–226.

Hirayama, Y., & Ronald, R. (2007). *Housing and social transition in Japan.* London: Routledge.

Hirayama, Y., & Ronald, R. (2008). Baby boomers, baby busters and the lost generation: Generational fractures in Japan's homeowner society. *Urban and Policy Review, 26*(4).

Hoekstra, J.S.C.M. (2003). Housing and the welfare state in the Netherlands: An application of Esping-Andersen's Typology. *Housing Theory and Society, 20*(2), 58–71.

Holliday, I. (2000). Productivist welfare capitalism: Social policy in East Asia. *Political Studies, 48,* 706–723.

Holzmann, R., & Hinz, R. (2005). *Old-age income support in the 21st century: An international perspective on pension systems and reform.* Washington, DC: The World Bank.

Hurd, M. (1990). Research on the elderly: economic status, retirement and consumption and saving, Journal of Economic Literature, 28, 565–637.

Izuhara, M. (2007). Turning stock into cash flow: Strategies using housing assets in an ageing society. In Y. Hirayama & R. Ronald (Eds.), *Housing and social transition in Japan.* London: Routledge.

Japelli, T., & Pistaferri, L. (2002). *Incentives to borrow and the demand for mortgage debt: An analysis of tax reforms* (CSEF Working Paper 90). Salerna: University of Salerna.

Johnson, C. A. (1982). *Miti and the Japanese miracle: The growth of industrial policy, 1925–1975.* Tokyo: Charles E. Tuttle.

Johnson, Norman. (1999). Mixed economies of welfare: A comparative approach, Prentice Hall Europe

Jones, C. (1993). The pacific challenge: Confucian welfare states. In C. Jones (Ed.), *New perspectives on the welfare state in Europe.* London: Routledge.

Kasza, G. J. (2006). *One world of welfare: Japan in comparative perspective.* Ithaca: Cornell University Press.

Kato, J. (1996). Review article: Institutions and rationality in Politics' three varieties of Neo-institutionalists'. *British Journal of Political Science, 27,* 553–582.

Katsura, H. M., & Romanik, C. T. (2002). *Ensuring access to essential services: Demand-side housing subsidies.* Social Protection Unit, Human Development Network, The World Bank.

Kemeny, J. (1980). Home ownership and privatisation. *International Journal of Urban and Regional Research, 4*(3), 372–388.

Kemeny, J. (2001). Comparative housing and welfare: Theorising the relationship. *Journal of Housing and the Built Environment, 16*(1), 53–70.

Kinsella, K., & Phillips, D. (2005). Global aging: The challenge of success. *Population Bulletin, 60*(1). Washington, DC: Population Reference Bureau.

Kluyev, V., & Mills, P. (2007). Is housing wealth an ATM? The relationship between household wealth, home equity withdrawal, and saving rates. *IMF Staff Papers, 54,* 539–561. doi:10.1057/palgrave.imfsp.9450018.

KNHC (Korean National Housing Corporation). (2009). *Yearbook on housing and urban statistics.* Seoul: KNHC.

Kurz, K., & Blossfeld, H.-P. (2004). *Home ownership and social inequality in comparative perspective.* Stanford: Stanford University Press.

Lefebure, S., Mangeleer, J., & Van den Bosch, K. (2006). *Elderly prosperity and homeownership in the European Union: New evidence from the SHARE data.* Paper to the 29th General Conference of the International Association for Research in Income and Wealth.

Lim, K. L. (2001). Implications of Singapore's CPF scheme on consumption choices and retirement. *Pacific Economic Review, 6,* 361–382.

Lusardi, A. (2000). *Saving for retirement: The importance of planning, research dialogue* (Issue No. 66). New York: TIAA-CREF Institute.

Malmberg, B. (2007). Demography and social welfare. *Journal of International Social Welfare, 16*, S21–S34.

Malpass, P. (2008). Housing and the new welfare state: Wobbly pillar or cornerstone? *Housing Studies, 23*(1), 1–19.

Mandic, S. (2012). Home ownership in post-socialist countries: Between macro economy and micro structures of welfare provision. In R. Ronald & M. Elsinga (Eds.), *Beyond home ownership: Housing welfare and society*. Oxford: Routledge.

Mandic, S., & Clapham, D. (1996). The meaning of home ownership in the transition from socialism: The example of Slovenia. *Urban Studies, 33*, 83–97.

McCarthy, D. (2004). *Household Portfolio allocation: A review of the literature*. Prepared for presentation at the Tokyo, Japan, February 2004 International Forum organised by the ESRI, Cabinet Office, Government of Japan, Japan.

McCarthy, D., Mitchell, O. S., & Piggott, J. (2002). Asset rich and cash poor: Retirement provision and housing policy in Singapore. *Journal of pension Economics and Finance, 1*, 197–222.

McKay, S. (2002) The savings gateway: "asset-based welfare" in action? *Benefits, 10*(2), 141–145.

Metropolitan Research Institute. (1996). *Regional housing indicators database in the transitional countries of Central and Eastern Europe: Indicators programme, monitoring human settlements*. Budapest.

Mulder, C. (2007). The family context and residential choice: A challenge for new research. *Population, Space and Place, 13*, 265–278.

Norris, M., & Shiels, P. (2004). *Housing developments in European countries, synthesis report*. Dublin: Department of the Environment, Heritage and Local Government, The Stationery Office.

North, D. (1991). Economic performance through time. *The American Economic Review, 84*(3), 359–368.

OECD. (2004, June). Housing markets, wealth and the life cycle. Paris: OECD Economic Outlook.

OECD. (2011). *Housing and the economy: Policies for renovation*. Paris: OECD.

Overton, L., & Doling, J. (2010). The market in reverse mortgages: Who uses them and for what reason? *Hypostat 2009*. Brussels: European Mortgage Federation.

Oxley, M. (2000). *The future of social housing learning from Europe*. London: Institute for Public Policy Research.

Oxley, M., & Smith, J. (1996). *Housing policy and rented housing in Europe*. London: E&FN Spon, an Imprint of Chapman and Hall.

Park, S. Y. (2007). The state of housing policy in Korea. In R. Groves, A. Murie, & C. Watson (Eds.), *Housing and the new welfare state: Examples from East Asia and Europe*. Aldershot: Ashgate.

Park, B.-H., & Lee, H.-O. (2007). A comparative study on housing welfare policies for the elderly between Korea and Japan – Focused on the elderly who can live independently. *Journal of Asian Public Policy, 1*(1), 90–103.

Peng, I., & Wong, J. (2004, September 2–4). *Growing out of the developmental state: East Asian welfare reform in the 1990s*. RC19 annual conference: Welfare state restructuring: Processes and social outcomes, Paris.

Pesaran, M. H., Shin, Y., & Smith, R. P. (1999). Pooled mean group estimation of dynamic heterogeneous panels. *Journal of the American Statistical Association, 94*, 621–634.

Phang, S. Y. (2007). The Singapore model of housing and the welfare state. In R. Groves, A. Murie, & C. Watson (Eds.), *Housing and the new welfare state: Examples from East Asia and Europe*. Aldershot: Ashgate.

Pierson, P. (2001). *The new politics of the welfare state*. Oxford: Oxford University Press.

Pierson, P. (2002). Coping with permanent austerity: Welfare state restructuring in affluent democracies. *Revue Française de Sociologie, 43*(2), 369–406.

Poggio, T. (2008). The intergenerational transmission of homeownership and the reproduction of the familiaristic welfare regime. In C. Saraceno (Ed.), *Families, ageing and social policy: Intergenerational solidarity in European welfare states* (pp. 59–87). Cheltenham/Northampton: Edward Elgar.

Poggio, T. (2012). The housing pillar of the Mediterranean welfare regime: Relations between home ownership and other dimensions of welfare. In R. Ronald & M. Elsinga (Eds.), *Beyond home ownership: Housing welfare and society*. Oxford: Routledge.

Priemus, H., & Boelhouwer, P. (1999). Social housing finance in Europe trends and opportunities. *Urban Studies, 36*(4), 633–646.

Regan, S., & Paxton, W. (2001). *Asset-based welfare: International experiences*. London: IPPR.

Reifner, U., Clerc-Renaud, S., Pérez-Carrillo, E., Tiffe, A., & Knobloch, M. (2009). *Study on equity release schemes in the EU* (Project No. MARKT/2007/23/H). Hamburg: Institut für Finanzdienstleistungen E.V.

Ritakallio, V.-M. (2003). The importance of housing costs in cross-national comparisons of welfare (state) outcomes. *International Social Security Review, 56*(2), 81–101.

Rohe, W. M., Von Shandt, S., & McCarthy, G. (2002). Home ownership and access to opportunity. *Housing Studies, 17*(1), 51–61.

Ronald, R. (2007). Comparing homeowner societies: Can we construct an east-west model? *Housing Studies, 22*(4), 473–493.

Ronald, R., & Chiu, R. H. L. (2010). Changing housing policy landscapes in Asia Pacific. *International Journal of Housing Policy, 10*(3), 223–231.

Ronald, R., & Doling, J. (2010). Shifting East Asian approaches to homeownership and the housing welfare pillar. *International Journal of Housing Policy, 10*(3).

Ronald, R., & Hirayama, Y. (2006). Housing commodities, context and meaning: Transformations in Japan's urban condominium sector. *Urban Studies, 43*(13), 2467–2483.

Ronald, R., & Jin, M. Y. (2010). Home ownership in South Korea: Examining sector underdevelopment. *Urban Studies, 47*(10).

Rouwendal, J. (2009). Housing wealth and household portfolios in an ageing society. *De Economist, 157*(1), 109–131.

Rowlingson, K. (2006). 'Living poor to die rich'? Or 'spending the kids' inheritance'? Attitudes to assets and inheritance in later life. *Journal of Social Policy, 35*(2), 175–192.

Rowlingson, K., & McKay, S. (2007). *Attitudes to inheritance in Britain*. York: Rowntree Foundation.

Rowlingson, K., Whyley, C., & Warren, T. (1999). *Income, Wealth and the Lifecycle*. York: Joseph Rowntree Foundation. Available at: www.jrf.org.uk/publications/income-wealth-and-lifecycle

Ruonavaara, H. (1993). Types and forms of housing tenure: Towards solving the comparison/ translation problem. *Housing Theory and Society, 10*(1), 3–20.

Sato, I. (2007). Welfare regime theories and the Japanese housing system. In Y. Hirayama & R. Ronald (Eds.), *Housing and social transition in Japan*. London: Routledge.

Scanlon, K., & Whitehead, C. (2007). *Social housing in Europe*. London: London School of Economics.

Scanlon, K., Lunde, J., & Whitehead, C. (2008). Mortgage product innovation in advanced economies: More choice, more risk. *European Journal of Housing Policy, 8*(2), 109–131.

Schmidt, S. (1989). Convergence theory, labour movements and corporatism: The case of housing. *Scandinavian Housing and Planning Research, 6*(2), 83–101.

SEQUAL. (2008). *"It's on the house" A consumer study into the attitude and perceptions of Australians aged over 60 years*. Sydney: Senior Australian Equity Release Association of Lenders.

Sherraden, M. (1991). *Assets and the poor: A new American welfare policy*. Armonk: M.E. Sharpe.

Sherraden, M. (2003). Assets and the social investment state. In W. Pxton (Ed.), *Equal shares: Building a progressive and coherent asset based welfare policy* (pp. 28–41). London: IPPR.

Shinkawa, T. (2005). The politics of pension reform in Japan: Institutional legacies, credit claiming and blame avoidance. In G. Bonoli & T. Shinkawa (Eds.), *Ageing and pension reform around the world: Evidence from eleven countries*. Cheltenham: Edwin Elgar.

Smelser, N. (2003). On comparative analysis, interdisciplinarity and internationalization in sociology. *International Sociology, 18*(4), 643–657.

Sobotka, T. (2008, December 2). *The rising importance of migration for births and fertility trends in Europe*. International conference: Effects of migration on population structures in Europe, VID-IIASA, Vienna.

Starke, P., Obinger, H., & Castles, F. (2008). Convergence towards where: In what ways, if any, are welfare states becoming more similar? *Journal of European Public Policy, 15*(7), 975–1000.

Stephens, M. (2003). Globalisation and housing finance systems in advanced and transitional economies. *Urban Studies, 40*(5–6), 1011–1026.

Stone, M. (2006). A housing affordability standard for the UK. *Housing Studies, 21*(4), 453–476.

Struyk, R. (Ed.). (1996). *Economic restructuring of the former soviet block: The case of housing*. Washington, DC: Urban Institute Press.

Tatsiramos, K. (2006). *Residential mobility and housing adjustment of older households in Europe* (Discussion Paper No. 2435). Bonn: Forscchunginstitut zur Zukunft der Arbeit (IZA).

The Gallup Organization. (2009). Survey on income in old age.

Torgerson, U. (1987). Housing: The wobbly pillar under the welfare state. In B. Turner, J. Kemeny, & L. Lundqvist (Eds.), *Between state and market: Housing in the post industrial era* (pp. 116–127). Gavle: Almqvist & Wiksell International.

Toussaint, J. (2011). *Housing wealth in retirement strategies: Towards understanding and new hypotheses*. Amsterdam: IOS Press.

Tridico, P. (2004). *Institutional change and economic performance in transition economics: The case of Poland* (Sussex Working Paper). Brighton: Sussex University.

U.N. (2005). http://www.un.org/esa/population/publications/WUP2005/2005wup.htm. Accessed 1 June 2010.

UNECE. (2006). *Guidelines on social housing: Principles and examples*. New York and Geneva: United Nations.

Van der Hoek, M. P., & Radloff, S. E. (2007). *Taxing owner-occupied housing: Comparing the Netherlands to other European Union countries* (MPRA Paper 5876). Munich: University Library of Munich.

Venti, S. F., & Wise, D. A. (2001). *Aging and housing equity: Another look* (NBER Working Paper w8608). Cambridge, MA: NBER.

Watson, M. (2009). Planning for the future if asset-based welfare? New labour, financialized economic agency and the housing market. *Planning Practice and Research, 24*(11), 41–56.

Weiss, L. (2003). Guiding globalisation in East Asia: New roles for old developmental states. In L. Weiss (Ed.), *States in the global economy: Bringing domestic institutions back in* (pp. 245–270). Cambridge: Cambridge University Press.

White, Gordon and Goodman, Roger (1998) Welfare Orientalism and the Search for and East Asian Welfare Model, in: Roger Goodman, Gordon White and Huck-ju Kwon (eds), The East Asian Welfare Model: Welfare Orientalism and the State, London: Routledge, pp 3–24.

Whitehouse, E. (2007). *Pensions panorama: Retirement pension systems in 53 countries*. Washington, DC: The World Bank.

Wolswijk, G. (2008). *Fiscal aspects of housing in Europe*. ECB, DG-Economics, http://www.oenb.at/de/img/guidowolswijk_tcm14-89925.pdf. Accessed on 10 Feb 2010.

World Bank. (1994). *Averting the old age crisis: Policies to protect the old and promote growth*. Washington, DC: World Bank Publications.

Yap, M. T. (2002). *Employment insurance: A safety net for the unemployed,* Institute of Policy Studies. Report Prepared for the Remaking Singapore Committee.

Index

J. Doling and M. Elsinga, *Demographic Change and Housing Wealth: Homeowners,*
Pensions and Asset-based Welfare in Europe, DOI 10.1007/978-94-007-4384-7,
© Springer Science+Business Media Dordrecht 2013

Printed by Printforce, the Netherlands